YELLOW FEVER
BLACK GODDESS

YELLOW FEVER
BLACK GODDESS

The Coevolution of People and Plagues

CHRISTOPHER WILLS

HELIX BOOKS

PERSEUS PUBLISHING

Cambridge, Massachusetts

Library of Congress Cataloging-in-Publication Data
Wills, Christopher.
   [Plagues]
   Yellow fever, black goddess : the coevolution of people and plagues / Christopher Wills
         p.   cm.
   "First published in the United Kingdom by HarperCollins Publishers under the title : Plagues : their origins, history, and future" - T. p. verso
   Includes bibliographical references and index.
   ISBN 0-201-44235-3 (alk. paper)
   ISBN 0-201-32818-6 (pbk.)
   1. Epidemics – History. I. Title.
RA649.W55   1996
614.4'9—dc20   96-23934
                                                   CIP

This work was first published in the United Kingdom by HarperCollins*Publishers* under the title *Plagues: Their Origins, History and Future.*

Perseus Publishing is a member of Perseus Books Group
Visit us on the World Wide Web at www.perseuspublishing.com

Cover design by Suzanne Heiser

4 5 6 7 8 9

To the memory of my grandfather,
E. H. ROBERTON,
who spun wonderful tales of India
and told me much more besides.

CONTENTS

FIGURES

AUTHOR'S NOTE

Even though it deals with the most serious of topics, this book has been a lot of fun to write. There is nothing more exciting than bringing together facts and observations gleaned from many different people and from many different parts of the world, and then finding that they make an astonishing and quite unexpected pattern that helps to illuminate the complexity of the living world. That is what happened in the course of writing this book, and I can only hope that I have managed to communicate the excitement of this intellectual chase to the reader.

The book is not, I should say at the outset, encyclopedic in scope. Our species is subject to so many diseases – some of which are plagues and some of which are far less obvious in their effects – that it would be unbearably tedious to examine them all in depth. I have chosen diseases that have, in striking ways, shaped our evolutionary history and that of the world around us. My apologies if I have left out a disease of particular interest to the reader!

In my travels, both of body and mind, I have been introduced to people and experiences quite outside the narrow world of academe that is my normal lurking place. The following people, from many parts of the world, have been most helpful in various phases of the project: Roy Anderson, Stuart Anstis, Francis Black, Jack Bradbury, Wilma Casanova de Caspía, Ted Case, Jim Comiskey, Vaughn Cooper, Francisco Dallmeier, K. K. Datta, David Dennis, Kitty Done, Norman Done, Renu Dutta, Dick Dutton, Stanley Falkow, Anthony Fauci, Joshua Fierer, Ken Gage, Bob Gallo, Uriel Garcia-Caceres, Jan Geliebter, Doug Green, Eduardo Groisman, Amiya K. Hati, Jonathan Howard, John Holland, Stephen Hubbell, Audrey Ishida, Mary Jesudason, T. Jacob John, Philip Gwyn Jones, Murli Krishna, Sheila Lukehart, Geeta Mehta, Stephen Morse, Juan P. Murillo, Gerald Myers, R. C. Panda, Curtis Patton, Margaret Perkins, Parukutty Pillai, V. S. Ramachandran, Jayanthi Ravi, V. S. Ravi, Miguel Rivera, Violeta Seminario de Rivera, George Riviere,

Milton Saier, K. B. Sharma, Robert Sinden, Cullen Stanley, Susan Swain, Geoffrey Targett, Glenys Thomson, Anne Marie Wills, Mark Wilson, Flossie Wong-Staal, Vishwanath Yemul, and Neville Yoon. They may not agree with everything I say here, but surely it is the author's privilege to go out on a limb occasionally. Errors of fact, of course, which I hope are few, are my responsibility entirely.

There is a series of readings arranged by chapter and page at the end of the book. And there is also a glossary to help with the unavoidable bits of terminology. Signs of these pedantic appurtenances are invisible in the text, however, and they can be ignored if the reader wishes.

PART ONE

The Anatomy of Plagues

1

The delicate balance between life and death

The Venice of the Amazon

Belén, a shantytown suburb of the city of Iquitos, lies on the banks of the Amazon River in eastern Peru, not far from the Brazilian border. Population pressure has forced it to expand in all directions, so that some of its palmetto-thatched huts have even been built out on pontoons that extend into a quiet eddy of the river itself.

The river flowing past Belén has gathered its water from a great network of rivers and streams that all snake their way eastward through the lowland rainforest. Already, during the season when it is at flood, the young Amazon is three kilometres wide, even though it still has a full three thousand kilometres to flow before it reaches the Atlantic. Indeed, it is quite possible for oceangoing ships to travel from the mouth of the Amazon to Iquitos and Belén and even beyond, having passed through almost the entire length of the Amazon basin.

In the nineteenth century, Iquitos was an important port for the vicious and exploitative rubber trade, but the city has since fallen into a kind of dreamlike tropical existence, with a little tourism, a good deal of logging and subsistence agriculture, some mining, and the inevitable drug trade.

The centre of Iquitos, which is all the tourists see, has a cheerfully raffish air. Crowds of girls swarm the streets, giggling whenever exotic visitors come into view. There is a pervasive throbbing mixture of tropical and Andean music. Children swarm everywhere, begging and offering small services. A permanent cloud of vultures adds a Graham Greene touch as they circle above the rusty tin roofs. But it is in the surrounding shantytowns like Belén that the real dynamic of this exploding Third World population centre becomes apparent.

Typically, the weather during the rainy season is hot, overcast and breathlessly humid. On just such a morning in March 1991, nurse Wilma Casanova de Caspía took a detour through the shantytown's crowded and cheerful street market on her way to work.

The market is not in the floating part of the town, but fills the streets of several nearby square blocks on dry land next to the shore. Displays of tropical fruit vie for the browser's attention with piles of meat and many different species of freshly caught fish, some exhibiting fearsome sets of teeth. Shy Indian women from the surrounding villages display mounds of peppers in a dazzling variety of colours, sizes and shapes. You can buy anything you might want in the market, from river turtles to astonishingly cheap Casio watches.

As Wilma strolled through the noise and bustle, she suddenly saw the people in front of her draw away. The shoppers had recoiled from a man in the middle of the street who was acting strangely. A fearful whisper ran through the crowd as it pulled back. The stricken man, his trousers soaked and glistening with fluid, weaved unsteadily, and at first she thought he was drunk. He staggered for one final time and collapsed almost at Wilma's feet. It was then she realized that the word the crowd was whispering, a word that suddenly filled the air, was *cholera*. She was seeing, for the first time in her life, the devastating effects of one of the most dreaded of all human diseases.

It is clear in retrospect that Belén is ideally situated to form the focus for an outbreak of cholera, and the wonder is that it did not happen sooner. An astonishing one hundred and twenty shantytowns like Belén are grouped around the city of Iquitos. They draw people like magnets from the two thousand tiny communities that dot the banks of the thousand or so rivers threading through the vast surrounding region of jungle that is known as Loreto. Most of the shantytowns consist of rows of huts lining unpaved roads that soon peter out in the forest. Each hut has a small cleared area behind it, no bigger than a few square metres, with enough room for a few vegetables, chickens, and a pig or two. Shanties closest to the centre of town are connected by slender threads of electric wires. Many of these glitter with newness, testimony to the frantic efforts of the Iquitos city government as it tries to catch up with the city's ever-expanding periphery. But none of the huts has

FIG. I-I The shanties of Belén, stranded on the riverbank during the dry season

water, except for shallow wells scooped down a few feet to the water table below.

Belén, the most unusual of these shantytowns, is known to local promoters as the Venice of the Amazon. Like Venice it has plentiful supplies of both water and garbage, but there the resemblance ends. Belén's equivalent of Venetian *palazzi* are its strings of palmetto-thatched shanties, all without water and most without electricity. In the wet season these shanties float in the river and boats can ply the canal-like stretches between them. In the dry season, from June to October, when the river can drop several metres below its maximum level, most of the shanties are stranded in rows on the shore, surrounded by a festering sea of mud.

At the driest part of the year, a visitor must climb ten metres up the mud bank of the river to the street above, past rotting piles of offal scattered among the huts. The river, that essential *cloaca maxima*, eventually sweeps all this downstream, but at the time I visited Belén and talked to Wilma about her experiences with cholera the rains were two months away and the flies swarmed in great clouds.

Wilma told me how, when she was confronted with her first victim in the market, she shouted vainly for help. Even though she radiated authority and was dressed in her blue-and-white public health nurse's uniform, nobody moved among the fearful crowd that packed the market. Then a porter, a friend of the stricken man, came forward. He picked up the victim like a bundle of bananas and followed Wilma towards the nearby main street. Here they found a continuous traffic of *motocars*, motorcycles with narrow back seats covered by brightly-coloured little canopies. These tiny vehicles are free to swarm every-where in Iquitos and its shantytowns, since there is no connection by road to the rest of the world. Wilma and the porter hired two of them and took the victim to Iquitos' main hospital.

The emergency room personnel had no idea what to do. But this case of cholera was hardly a surprise. Beginning two months before, there had been a devastating outbreak of cholera on the Peruvian coast, making its appearance in several cities almost simultaneously. News of the epidemic had spurred the establishment of a cholera unit at the hospital, and Wilma immediately found the doctor in charge. He was familiar with the disease, and knew that cholera kills by dehydration as the body's fluids rush unimpeded through the intestinal wall. The victim, now comatose, was put on a bed in the cholera unit and four plastic bags containing a physiological salt solution were attached to him, one for each arm and each leg. Within a few hours he had recovered sufficiently to be able to recognize and smile at his rescuers.

Wilma learned the location of the man's village from his friend the porter, and was soon on her way there in a dugout canoe. In the village she found that the victim's wife and children had also become violently ill with diarrhoea during the day, although the symptoms were not as dramatic. She added packets containing a mixture of rehydration salts and glucose to water, and forced them to drink copious amounts. They too recovered quickly.

Cholera can strike with appalling suddenness. The man had been in perfect health when he left his family at five o'clock that morning to travel to the Belén market. By eight, when Wilma saw him, he was in such a state of violent dehydration brought on by the disease that one more hour without treatment would have produced irreversible brain and kidney damage.

The Walrus and the Carpenter

'I weep for you,' the Walrus said;
'I deeply sympathize.'
With sobs and tears he sorted out
Those of the largest size,
Holding his pocket-handkerchief
Before his streaming eyes.
'Oysters,' said the Carpenter,
'You've had a pleasant run!
Shall we be getting home again?'
But answer came there none –
And this was scarcely odd because
They'd eaten every one.
Lewis Carroll, *Through the Looking Glass* 1872

Iquitos has exploded suddenly like an ecological bomb in the middle of the Andean rainforest, a highly diverse and extremely stable ecosystem. We look at Iquitos and are hardly surprised that a plague of cholera broke out there. It is after all a truism that the plagues that afflict us do so because we have upset the balance of nature.

We have been upsetting this balance for a very long time. Fossil bones of our remotest ancestors found in East Africa provide evidence that those ancestors were already beginning to modify their environment. The discovery by Donald Johanson of the 'First Family', that astonishing collection of Australopithecines that has been dated to 3.2 million years ago, was remarkable for many things. Among them was the striking fact that no other animal bones were found among the remains of these hominids, so animals were not present when the disaster that killed them struck. This suggests that our ancestors were already finding ways to band together and live apart from the natural world, and to alter that world in order to make it safer for them.

Such alterations can be extensive. Waves of extinctions of large animals, and dramatic changes in entire ecosystems, have always accompanied our own spread across the planet. Fifty thousand years ago, newly-arrived Aboriginals changed the entire ecology of Australia when they began to burn off huge regions of that continent. Ten thousand years ago hunters from Asia, working their way slowly down the coast of North America, wiped out the last of the great herds of animals living in California's Central Valley. And the first humans to arrive on the island of Madagascar, a thousand years ago, soon drove to extinction the largest birds that ever lived, *Aepyornis*. All that we have left of this mighty bird is the legend of Sinbad's Roc in the *Thousand and One Nights*.

Unfortunately, upsetting the balance of nature just happens to be what our species has been selected to do well – although we hate to admit it. Like the Walrus in Lewis Carroll's poem, we shed hypocritical tears over the diminishing supply of oysters, while gulping them down as quickly as ever.

In this book I would like to explore with you a biologist's view of just how and why such disturbances of the ecological balance can give rise to plagues. Here I define plagues as great epidemic or pandemic disasters. These include not just the bubonic plague or Black Death with which you are already familiar, but a wide variety of diseases which given the right conditions can spread rapidly and cause high mortality.

The terms epidemic, pandemic and plague tend to be used rather loosely and to have overlapping meanings. Strictly speaking, an epidemic is confined in time and space – it is derived from the Greek for 'visit'. A pandemic is an epidemic of great extent, sweeping across whole continents or the entire planet. The influenza outbreak following World War One was a pandemic, and it could also justifiably be considered a plague – it was responsible for at least twenty million deaths. More recent disease outbreaks, while they can often travel swiftly around the world, usually cause far less mortality – they are epidemics or pandemics, but nobody would call them plagues. This is because we have largely learned to control the worst manifestations of such diseases. If a cholera outbreak as extensive as the 1991 outbreak in Peru had happened in 1850, it would have been a plague, because it would have resulted in mortality of up to fifty per cent. But the 1991 epidemic, although it was part of a larger pandemic of cholera that has been slowly spreading around the planet for decades, had a mortality rate of only one per cent. It was not a plague, thanks to our growing expertise in fighting the disease, though in an earlier time it could easily have been.

The plagues that result from ecological disturbance make up only a small part of the story that we will trace here. The majority of diseases are *endemic*, rather than epidemics or pandemics or plagues. They afflict every species of animal and plant, and it turns out that they have a very important role to play in the natural world. At the end of the book we will explore how these endemic diseases have contributed greatly to the diversity of life on our planet, including our own diversity.

A surprising amount of the evolution of living organisms has been driven by such diseases. And we will see that this is not a one way street. At the same time that many different host species have been driven in the direction of greater genetic diversity in order to survive their diseases, the disease organisms that afflict them have not stood still. They have themselves evolved towards greater diversity. Much of this evolution has happened in undisturbed ecosystems, like the rainforest that still surrounds Iquitos. The full dimensions of this remarkable story, combining ecology and evolution, are only just beginning to emerge.

As we explore the interaction between ecology and disease, we will also find that our own activities are not entirely to blame for creating plagues. Although we are the most effective disturbers of ecosystems who have ever lived, and also happen to be the first inhabitants of the planet who can write down the history of our diseases, we now know that plagues have afflicted most species of animals and plants for most of the time that life has existed on our planet.

The Fourth Horseman

We are not quite ready to examine this ecological story, however. First we must try to understand the dynamics of the diseases that afflict our own species. We have thrown up an astounding variety of defences against our diseases, particularly the endemic ones, and they have in turn responded in very complex ways. Of course, our defences against these endemic diseases can be overwhelmed by the disasters of plagues, for we cannot defend ourselves against the unexpected. We have evolved to be very good at defending ourselves against a multitude of disease organisms as long as each is rare, but it is as hard to adapt to surviving plagues as it is to adapt to surviving earthquakes or volcanic eruptions.

Surprisingly, the organisms that cause plagues are themselves often ill-adapted. We are lucky that, despite their apparently fearsome nature, these plague organisms have weaknesses that we can exploit.

They do not appear weak, of course, to people caught up in the disastrous events that spawn them. Although plagues and threats of plagues have always existed, they have never seemed so immediate as today, thanks to television.

In July 1994, over a million terrified Hutu refugees poured out of the mountainous land of Rwanda into the town of Goma and the surrounding countryside just across the border in eastern Zaire. They had been driven there by fear of the Tutsi, a physically similar but on average taller tribe who had provided the kings of the area for centuries, and who had jealousy hoarded much of Rwanda's political and social power. Earlier in the year the latest in a series of Hutu rebellions had released hatreds that had been building up during all that time. These culminated in massacres of many of the Tutsi. Soon, however, the Hutu rebellion fell apart as the better-armed and organized Tutsi struck back. Now immense numbers of the Hutu were fleeing what they fully expected would be a fearful revenge. The minority who were guilty of the atrocities blended easily into the stream of innocent refugees.

The world's television screens were suddenly filled with apocalyptic landscapes strewn with the limp figures of tens of thousands of exhausted refugees, starving and desperate for water. Members of the defeated Hutu army stalked the camps, threatening and stealing from the innocent and readily killing their own tribespeople if they resisted. The sky was darkened by ash from nearby Mount Nyiragongo, which had chosen this dreadful moment to erupt.

Relief agencies were overwhelmed. At their peak, the refugee areas near Goma and elsewhere may have held as many as two million people, with little food and no shelter. Half of the refugees were children, many of whom had been separated from their families. Any sort of sanitation was impossible—the nearby lake, the only source of water, was polluted by corpses and sewage. Even if tools had been available to dig latrines, the hard lava that lay just beneath the camps' thin soil would have made it impossible.

The first cases of cholera appeared within two or three days of the refugees' arrival. The severity of the outbreak increased by the hour, and soon at least 250 refugees were dying each day. The corpses, shrouded in pitiful rags, began to pile up unburied by the side of the road that led into the camp. Cholera was closely followed by *Shigella* dysentery. The world was stunned by the astonishing speed with which these diseases had appeared and spread through the camps. The threat of uncontrolled plagues finally jarred governments into action, belatedly reinforcing the efforts of *Médecins sans Frontières* and other private relief organizations.

Within weeks, the US Army managed to get water purification equipment into place. Clinics were set up to treat the thousands of wounded and starving and to administer rehydration therapy. The epidemics were brought under control with surprising speed, and the ragtag camps slowly began to shrink as people tentatively and fearfully drifted back to their towns and villages. The situation had gone from apocalyptic to merely desperate. The world's attention shifted elsewhere.

That elsewhere turned out to be India, and the disease that broke out there was a startling echo of the medieval world – bubonic plague.

The first probable appearances of plague in India are recorded in the ancient chronicle of the *Bhagvata Purana*, dating from as long ago as 1500 BC. In AD 1031 a new wave of plague swept in from Central Asia with the army of Sultan Mohammed, and in succeeding centuries Tamerlane and other conquerors from the north brought plague along with the other disasters of war. Plague's dreadful toll continued down to the present century. The worst outbreak in modern times took place in 1907. In that year, plague swept through India's cities and 1.3 million people died. But this was only one episode in wave after wave that swept over the subcontinent between 1898 and 1918, during which an astonishing total of twelve and a half million people lost their lives. This is a quarter of the number of people who died in all of Europe during the Black Death of 1348, and it is amazing to realize that this slaughter happened within the memory of people alive today.

The last really serious outbreaks of plague before its 1994 reappearance took place during the massive dislocation of populations that had been triggered by the partition of India and Pakistan in 1948. But by the 1970s, with only a few possible but unconfirmed cases being reported, it seemed that the disease had effectively disappeared. As time went on, India's National Institute for Communicable Diseases discontinued all but three of its plague surveillance units. Then came a series of unusual events.

The first happened on 30 September 1993, almost a year before the new appearance of plague. It took the form of a devastating earthquake of magnitude 6.4, centred near the town of Latur some 350 kilometres inland from Bombay. This earthquake was the most deadly ever to have been recorded in any part of the earth's crust far from known fault lines, killing more than 11,000 people. The disaster alerted Indian health authorities. They were very aware of the possibility that wild rats and other animals that harboured the plague bacillus could be displaced from their usual haunts, mixing with domestic rat and human populations and triggering an outbreak of plague.

The second event happened in the small town of Beed, over 100 kilometres from Latur. A local doctor reported the finding of numerous dead rats, known as a 'ratfall', and of flea bites afflicting the inhabitants of small villages near the town. This took place early in August 1994.

Then, starting on 26 August, villagers began to turn up in the local clinics with high fevers. Some had severely swollen lymph nodes in the armpits and groin. On 2 September, K. K. Datta, director of the National Institute of Communicable Diseases in Delhi, got a worried phone call from a health officer in the city of Poona, near Beed. Datta asked him for samples of blood sera from the patients. These were shipped to Delhi and tested over the next few days for antibodies against the plague bacillus *Yersinia pestis*. Seven out of ten of the samples appeared to be positive, though the test was a crude one dictated by the unavailability of more sophisticated reagents. Datta arranged for a team of entomologists and microbiologists to fly to the area.

The number of apparent cases in the villages near Beed continued to increase, eventually totalling 460. Although treatment with the antibiotic tetracycline was only begun several days after the appearance of the first case, there were no deaths. This is surprising because, unless plague is treated very soon after the infection is detected, mortality can be as high as fifty per cent. Even in the US, about fifteen per cent of plague cases die in spite of treatment. Further, there were no instances of the massive lung, brain and intestinal infections that almost always begin to appear as a plague outbreak advances.

Then the focus of attention suddenly shifted to the north, to Surat. This grim industrial city, the site of the first British settlement in India in the early seventeenth century, is on the coast some 250 kilometres north of Bombay – and 400 kilometres from the outbreak in Beed. In August, Surat had been hit hard by monsoon rains which had left numerous piles of garbage and dead animals. On 19 September, people began to arrive in the city's hospitals with a frightening set of symptoms.

They had high fevers, had great difficulty breathing, were racked with coughs, and showed blood in their sputum. Could this be plague as well?

Yes, it could. Plague comes in three forms, depending on how widely the bacterial infection has spread in the body. In the classical bubonic plague the bacteria accumulate in the lymph nodes, with spill over into the bloodstream. The swollen lymph nodes (buboes) are a sign of the frantic response of the patient's overwhelmed immune system. This form of plague can only be spread by fleas, which transfer the infected blood to a new victim.

In septicaemic plague, large numbers of the plague bacilli swarm in the bloodstream and can escape into other tissues such as the lungs and the brain. And in pneumonic plague, the most infectious type, the bacteria multiply swiftly in the lungs and are broadcast through the air in tiny droplets. If plague reaches the pneumonic stage it can spread like wildfire, particularly when people are crowded together, for it no longer requires rats or fleas. This is the most dangerous and frightening manifestation of a plague outbreak. And, in contrast to the cases that had been seen in the Beed district, it appeared that those in Surat were pneumonic.

Public health authorities in Surat, caught unprepared by this new outbreak, did not at first admit that there was anything wrong. Then, as the number of cases mounted, they overreacted. Radio and television broadcasts announced the presence of plague, and trucks with loud-speakers blared the same message in the poorer parts of the city. Panic ensued. Over the next three or four days Surat essentially shut down. Bus and train stations were besieged by frantic people trying to leave. It is unclear how many succeeded. Early reports suggested that half a million people, one-third of the city's population, managed to flee, but this seems unlikely. India's inadequate train system could never have moved that many people in such a short time. Indeed, train records showed that only about fifteen per cent more tickets than usual were sold during the time of the panic. Other people who were able to do so left by bus and car, and among those who fled, unfortunately, were many of the doctors of Surat.

Although the numbers who fled the city were probably not in the hundreds of thousands, there is no doubt that thousands of people from Surat fanned out over India, going for the most part to the homes of relatives. At least fifteen hundred of them are known to have travelled as far as West Bengal, on the other side of the Indian peninsula. After four days, there were some government attempts to seal off the roads and railway lines, though people fleeing by car found no sign of the roadblocks that had been announced. In short, there was utter confusion and wild exaggeration, and the panic quickly spread to the rest of India and the rest of the world.

At the panic's height all but two international airlines shut down flights to India. The Gulf states, among India's most important trading partners, banned imports of Indian foodstuffs. At airports around the world, passengers arriving from India were scrutinized for signs of fever. In Toronto, airport workers donned gloves and masks. Eleven fever cases arriving in New York from India were quarantined, but all turned out to have something else wrong with them. (Astonishingly, four had malaria and one had typhoid fever, giving a snapshot of the kinds of active diseases that are ferried every day around the planet by the world's airlines.)

Reports of plague cases in Bombay and Delhi set off further panics. Hospitals were flooded with people, most of whom were found to be suffering from minor ailments and were quickly discharged. The eventual response of the government, while belated, was dramatic. By 19 September, the area within a fifteen-kilometre radius around Beed had been sprayed with hundreds of tons of DDT and other insecticides. Bombay's Haffkine Institute had 800,000 tetracycline tablets on hand that were immediately available for transport to Surat, and other sources were quickly found. Within days of the Surat outbreak an astonishing eight million tablets had arrived by airlift, and another two million were sent to Beed.

These heroic measures quickly contained the disease, but left unanswered the question of whether it had really been plague. When I visited India at the beginning of 1995, controversy still raged, but one thing seems certain. No plague-like outbreak could have withstood the deluge of tetracycline and DDT that descended on it. In this case, the power of twentieth-century technology had proved overwhelming.

Navajo wisdom

Technology can not only stamp out plagues in their early stages, but can also permit new diseases to be tracked down and understood with unprecedented speed. The Four Corners virus that recently appeared in the US is a vivid example. The disease's existence was first suspected in the middle of May 1993. Two members of the same New Mexico family, living not far from the Four Corners area where the boundaries of four states touch, died of acute respiratory distress within five days of each other. The apparent coincidence alerted local public health workers, who quickly examined necropsy tissues from the victims' lungs. They were able to rule out any of the obvious bacteria and viruses.

Samples of lung tissue from the victims arrived at Atlanta's Centers for Disease Control on 31 May. By 2 June, sensitive immunological assays had yielded a very unusual result. The disease agent in the cells was a virus, which was expected, but it turned out not to be a relative of the known viruses that can cause pneumonia. Instead, it was related to a group called hantaviruses. These are widespread in Asia and Europe, and are responsible for diseases that cause haemorrhaging in the kidneys. The hantavirus connection was unexpected, since the Four Corners virus clearly affected the lungs. Its symptoms were completely different from those of the other hantaviruses.

The next step was to use a powerful molecular technique, the polymerase chain reaction, widely known by its abbreviation PCR. Tiny amounts of DNA that had been isolated from the diseased tissues of the infected individuals were added to PCR reaction tubes. Somewhere in this mixture, swamped by the enormous amounts of DNA from the cells of the victims, were a few DNA molecules that had been copied by their cellular machinery from the RNA of the virus. The trick was to *amplify* this tiny amount of DNA, to make enormous quantities of it that would enable the message that it carried to be read.

The PCR reaction itself is extremely simple. Mix short stretches of DNA called primers with a bit of the DNA mixture you want to examine, and add a heat-resistant enzyme that can build new DNA strands using the primers as starting points. Heat and cool this solution repeatedly according to a precise schedule, and after 20 or 30 cycles you will have enormous quantities of the piece of gene that lies between the primers (Figure 1-2 shows how this is done).

The result was enormously amplified quantities of a chunk of the hantavirus DNA. These pieces were amplified on 8 June, and the sequence of their bases was determined by 9 June, less than a month from the time that the disease had initially been detected.

Once the DNA was available, tissue samples from wild animals that inhabit the region were then rapidly screened for the presence of these sequences, again using PCR. The virus turned out not to be a human disease at all. Rather, it was a natural pathogen of a number of different rodent species, and was particularly common in the charming little white-footed deermouse *Peromyscus maniculatus*.

In a dazzling tour de force of biological detective work, it was soon shown that the Four Corners virus had lived in these rodent populations for a long time, for it was made up of a number of different strains that over many thousands of years had acquired slightly different genetic sequences. The virus, which was found to be responsible for a burst of subsequent cases throughout the western US, turned out to be an old enemy, not a new one. Indeed, it was probably responsible for some of the rare cases of fatal respiratory disease that over the years had occasionally mystified doctors in the rural Southwest.

It is ironic that if people had paid attention to local Navajo tradition, which speaks of mice as agents of disease and therefore creatures to be avoided, the outbreak might not have happened. But the ever-increasing numbers of people in the area have probably made this mini-plague inevitable.

The Four Corners virus cannot, at least in its present form, be transmitted from one human to another. As a result, the route of infection can easily be blocked. People throughout the region were quickly alerted to the dangers of coming in contact with wild mice and their droppings, and the number of reported cases diminished just as quickly.

The speed with which this disease was tracked down, understood, and, at least in the short term, held at bay in the affected population, is remarkable. Luckily, the Four Corners virus is a relatively easy one to control, and since related viruses were already known it could be diagnosed rapidly.

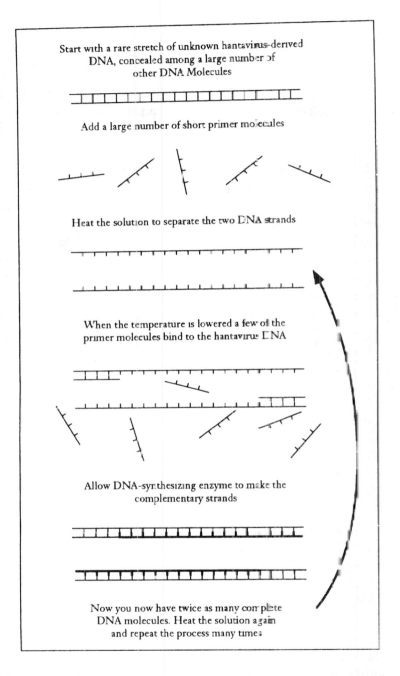

Start with a rare stretch of unknown hantavirus-derived DNA, concealed among a large number of other DNA Molecules

Add a large number of short primer molecules

Heat the solution to separate the two DNA strands

When the temperature is lowered a few of the primer molecules bind to the hantavirus DNA

Allow DNA-synthesizing enzyme to make the complementary strands

Now you now have twice as many complete DNA molecules. Heat the solution again and repeat the process many times

FIG. 1-2 How the PCR technique can be used to multiply small quantities of DNA into immense numbers of copies.

A cool look at the hot zone

Plagues of various kinds seem to be appearing everywhere. In southern Russia and the Crimea, the end of the hot summer of 1994 brought a wave of cholera cases, with a dozen deaths. And in the depressed and financially ruined cities of Moscow and St Petersburg to the north the resurgent enemy is now diphtheria, a childhood illness that had been thought to be a thing of the past. There were almost 50,000 cases of this choking disease in the first half of 1994, up from 1200 in 1990 and 4000 in 1992. Most were children, and about one in twenty of the victims died.

Plagues or threatened plagues of these old diseases are being matched by a horrifying variety of new perils. In Africa, the widespread invasion of the rainforest by the exploding human population has disturbed nests of haemorrhagic viruses with names that sound like a drumbeat of doom: Lassa, Ebola, Marburg. In South America, three other viruses called Junin, Machupo and Guanarito, all related to Lassa, have recently emerged, again apparently as a result of human expansion into the forests. They have caused small but highly lethal clusters of cases in Argentina, Bolivia and Venezuela.

No continent and no group of people seem to be immune to such outbreaks. One of the fatalities caused by the recent outbreak of the Four Corners virus was a graduate student in my own department at the University of California, San Diego. She came in contact with it quite unknowingly when she was carrying out field observations on birds in California's Sierra Nevada mountains. She was exposed to the virus while living in a cabin that had been deserted for much of the year and that had become infested with fieldmice.

Luckily, so far, the Four Corners virus and most of the other so-called emerging viruses – with one notable exception – have not spread widely. This is because they cannot be transmitted easily from one human to another.

The worst of the emerging viruses, of course, and the greatest plague that is facing us at the end of the twentieth century, is HIV, the AIDS virus. Unlike most of the other emerging viruses it is transmitted from one human to another, though not easily, and its rate of spread continues to increase, particularly in the Third World.

Especially in sub-Saharan Africa, where it is very widespread, AIDS puts new strains on fragile societies. The very knowledge of the apparent inevitability of the disease has plunged millions of people into numb despair. In that region AIDS is only one part of a grim cycle involving other diseases, poverty, ignorance and warfare, so that cause and effect can no longer be separated. But there is no doubt that its effects on social structures have contributed to the killings in Rwanda and the waves of ethnic slaughter that have taken place elsewhere in Africa.

Although a larger fraction of our species is living disease-free lives than ever before, thanks to advances in medicine and public health, the perception in the public mind is that plagues, and the threat of plagues, are ever-present. Recent books such as *The Hot Zone* and *The Coming Plague* have painted a truly terrifying picture of these plagues, and of their possible future successors. At any moment, the authors suggest, emerging haemorrhagic viruses might burst forth, be carried around the planet by airline passengers, and quickly turn us all – or at least all but a few resistant survivors – into piles of mush.

What are the odds that this will happen? What properties might such a plague need in order to spread so uncontrollably? And what are we doing to ourselves and our planet that might bring on such an apocalyptic event? Where do plagues come from, anyway? Why do they happen?

All these diseases are part of a much larger picture. We must understand that picture if we are to face up to them and take the steps needed to conquer them. We must also confront fully the extent to which our own activities trigger the outbreaks we call plagues.

Waves of disease have struck the human race many times before. Indeed, many of these waves were a good deal worse than the current surge, since nothing could be done about curing the diseases or about immunizing the people who were likely to be infected. They struck with particularly devastating results when conditions were changing rapidly. For example, populations were undergoing dramatic increases during the Age of Exploration in the fifteenth and sixteenth centuries, just as they are at the present time. New technology allowed people to move long distances with greater ease. Then as now, there was widespread and continual warfare. The invasion of unexplored parts of the world was exposing whole groups of people to new diseases. All these human activities, then as now, were accompanied by disease outbreaks.

But is there something unusual in the current outbreaks? Is it possible that they foreshadow the end of our species?

Of course, our species was not wiped out by its diseases in the past, and it will not be wiped out by them now. The world is not coming to an end, although there is no doubt that we, both as a species and as individuals, are in for some very unpleasant and dangerous times. This is because there are things about the current wave of plagues that are new. Some of them are obvious and some are not so apparent.

First, the obvious. The human race is expanding in numbers at an unprecedented rate. We are spreading into the last unexplored corners of our world, and in consequence we are flushing new diseases from their hiding places. When this expansion is combined with the devastation of war, as in Rwanda, or with economic upheaval, as in Russia, the mix can be particularly deadly. This has always been true, but because of new medical technology and instant communications it has never been so thoroughly documented as now.

It is less obvious that our sheer numbers provide an unprecedented opportunity for disease organisms to evolve. In most parts of the planet, so far as these organisms are concerned, we are the largest meal around. As our population multiplies, so do those of the disease organisms that attack us. The human population explosion is also a pathogen population explosion, with all the evolutionary consequences that this implies.

Mutations are the feedstock of evolutionary change, and in theory they may happen at any time. The ultimate epidemiological nightmare, always hanging over our heads like the sword of Damocles, is that a disease like AIDS or Ebola will suddenly become highly infectious through mutation. Perhaps these viruses will acquire the ability to travel through the air, or begin to spread with the aid of mosquitoes and other biting insects. There is no doubt that such a horrific event is made much more probable by the presence of huge populations of these pathogens, since the larger the population the more likely it becomes that a rare mutation or some other kind of rare genetic event will occur somewhere within it.

More mundanely, pathogens have often borrowed from their relatives genes that can increase their virulence. In Washington State, in January 1993, a consignment of hamburger meat arrived in the kitchens of the Jack-in-the-Box fast-food restaurant chain. It happened to be bacterially contaminated, and was cooked less thoroughly than it should have been. The result was an outbreak of *Escherichia coli* food poisoning which caused three deaths and more than 500 cases of severe illness. The outbreak was caused by a variant strain, called O157 H7, of this usually mild-mannered bacterium. This strain carries, among a nasty assortment of other genes, a gene for a toxin that is shared widely by a rogues' gallery of *E. coli*'s relatives – *Shigella Yersinia, Campylobacter,* and *Salmonella*. Although most strains of these various bacteria do not have the toxin gene and live relatively benignly in our guts or in the soil, there are a few variant strains of each species that have borrowed this gene and other dangerous genes in the past. Such borrowings happen rarely, but the likelihood is increasing that strains carrying these rare combinations will be found somewhere in the huge mixed populations of bacteria that our activities are creating.

A more subtle problem has grown out of our very success at fighting disease. Many of the 'easy' diseases, those that are readily susceptible to antibiotics or that we can easily be immunized against, have now been conquered or driven to very low levels. The medical historian Mirko Grmek has suggested that while this has lengthened our lifespans it has also provided an opportunity for rarer, less tractable diseases. AIDS, which takes an average of ten years to kill its victims in industrialized societies, might not have spread so effectively in a world in which people who had been infected with the virus would usually die of something else first. Grmek's example of AIDS may be poorly chosen, since people who have just acquired the virus tend to be the most infective. But it is certainly true that relatively intractable diseases are posing an ever greater danger in many societies.

Nowhere is this picture changing more rapidly than in India. A lifetime ago, bubonic plague was a terrible cause of mortality in that teeming country. These days, the greatest danger is far more subtle. As one wanders through the crowded streets of an Indian city or town, it is sobering to realize that over a third of the inhabitants are infected with tuberculosis. Only a few per cent will develop overt signs of the disease during their lifetimes, but these cases still number in the millions. Treatment is difficult, expensive, and lengthy, involving three or even four drugs, taken daily, for a period of six to eight months. Although the cure rate for those who complete this demanding regimen is high, only about half the patients lucky enough to begin the treatment successfully finish it. When the treatment is stopped partway through, the surviving bacilli are those that are the most drug-resistant. As a consequence, drug-resistant strains of the TB bacillus are spreading, making new cases even more difficult to cure.

The most encouraging new factor in this litany of disasters is that we are beginning to understand plagues, and to understand them at every level. We cannot give final answers to the question of the nature of future epidemics, but we can formulate the question itself much more sensibly. Epidemiologists have a much clearer idea of how diseases are spread, and can determine the points in their life cycles at which they are best attacked. Our understanding of the cells that make up our bodies and of how they are attacked by pathogens has also grown enormously. Public health officials know the steps that should be taken to limit plagues. Indeed, they are frustrated primarily by a lack of resources and by the prejudice, ignorance, and fear of the public and of their elected officials, particularly when they are faced with an upsurge of sexually transmitted diseases.

But whether we can control these diseases or not, we must still confront the most unnerving question of all. What place do plagues and endemic diseases have in the natural world?

One of my most searing memories is of being surrounded on a street in Hyderabad by a crowd of lepers, ranging from young people to withered ancients, extending their fingerless hands towards me for alms. The faces of many of them had become collapsed and distorted from the ravages of the disease.

Not long before, I had visited a sheltered workshop in the grounds of a hospital in southern India, where similarly crippled lepers were being taught to do simple tasks. In India, leprosy is so prevalent that such programmes are hopelessly inadequate – the leprous beggars who crowded around me on the Hyderabad street had been unable to find a place in them. How is it that diseases as loathsome as leprosy or as deadly as bubonic plague have made a place for themselves in the living world?

Generators of diversity

Some years ago immunologist Mel Cohn, trying to explain the complexity of the human immune system, drew a box on the blackboard labelled GOD. He explained, deadpan, that this stood for 'Generator of Diversity'.

Cohn's little joke leads us to a much deeper answer to the question of the existence of plagues and other diseases. They are generators of diversity. Exploring this idea will lead us far beyond our own species, into an examination of the web of ecological interactions that connects us to all the other species with which we share the planet. We will discover that plagues form a part, a small part, of a kind of evolutionary feedback loop that has actually helped to generate the great diversity of the living world.

This diversity is one that encompasses both diseases and their hosts. We are challenged with a bewildering variety of diseases – assorted worms, protozoa, bacteria, fungi and viruses that, unchecked, can roister in our bodies in prolific abundance. The same applies to other animals and to plants. One answer to why there are so many diseases is simply that there are so many species on which they can prey.

But this immediately raises another question – why should there be so many species of animals and plants in the first place? As we will see, at least part of the answer to this question is that they are so numerous because there are so many diseases.

Surely this reasoning is circular! Not quite. As we will see, particularly in the tropics, there are so many pathogens that no host species is capable of defending itself against them all. And at the same time no pathogen, confronted by all these potential host species, is sufficiently versatile to be able to attack all of them. This relationship between pathogens and hosts has actually helped to drive the evolution of the teeming complexity of tropical life, and to a lesser extent life in the temperate zones as well. In the process of becoming very different from each other, host species protect themselves from most of the pathogens that might otherwise be a threat. And because each pathogen must concentrate its efforts on overcoming the defences of a particular type of host, pathogens inevitably become more and more specialized, and thus they too become more and more different from each other. The two processes reinforce each other, and as we will see they have particularly left their mark on the pattern of diversity in the tropical rainforest.

Since our species originated in the tropics, we are both the victims and the beneficiaries of this diversity. We are victims because of all the misery and death caused by these diseases. And we are beneficiaries because, even within our own species, this host–parasite competition is operating, helping to drive our own genetic diversity.

The competition between ourselves and our pathogens is manifest at every level, and its effects have even dramatically shaped the families of protein molecules that make up our cells. We now know enough about these proteins to measure the consequences of this fierce competition directly. Many different proteins have been sequenced – that is to say, the precise sequence of the smaller molecules called amino acids that are strung together to make up their structure has been determined. Recently, Philip Murphy of the US National Institute of Allergy and Infectious Diseases scanned through the rapidly growing database of such sequenced proteins, looking for those that are found both in humans and in rodents. He arranged these sequences in pairs, one from a human and one from a rodent. Each of these pairs can be traced back to a common ancestral molecule which was probably present in the little insectivore from which humans, rats and mice are descended. The little insectivore lived before the dawn of the Age of Mammals, sixty-five million years ago.

Murphy measured how far these pairs of proteins had diverged from each other in the course of the separate evolution of humans and rodents (Figure 1-3). He found that proteins involved in defending the hosts against bacteria, viruses and other pathogens have diverged three times as quickly as proteins that are not involved in these defences. Something, he thought, must be driving all this accumulation of diversity in our defences. That something seemed very likely to be the very different sets of pathogens that have afflicted humans and rodents in the course of their separate evolution. For tens of millions of years our ancestors have been in a frantic race with their pathogens, trying to keep up as they evolve. At the same time, the ancestors of mice and rats have been in a race of their own, fighting off the continually shifting malign effects of their own sets of pathogens.

Such genetic variation has accumulated, not only between species as they have diverged from each other over millions of years, but within species as well. This within-species accumulation, too, has been driven by those furiously evolving pathogens. The immunologist Douglas Green and I have shown that even though our species is afflicted by a wide assortment of diseases, this diversity tends to spread the burden, so that it is not too onerous for our species to bear. This is because we are so different from each other that no species of pathogen can become too prevalent – the majority of us are resistant to it, even though a few of us may be susceptible. We have called this complex *modus vivendi* between ourselves and our numerous pathogens, for reasons that will become clear, *genetic herd-immunity*.

We must understand this balance between life and death, and how it can be upset, for the following terribly cogent reason.

Human activity not only disturbs ecosystems, it simplifies them. Even

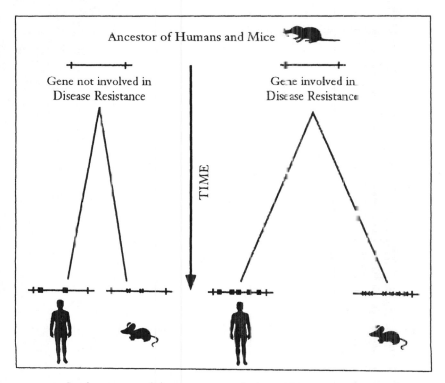

FIG. I-3 In the course of the separate evolution of humans and rodents, genes involved in disease resistance have diverged far more from each other than genes that are not involved.

though we come in daily contact with at least as many other living organisms as our remote ancestors did, the number of species is far fewer and the mix is very different. We have deliberately surrounded ourselves with creatures that are friendly to us, like dogs and cattle and wheat and corn. And we have also surrounded ourselves with other organisms that we have not meant to. Our activities have inadvertently helped these other organisms to multiply into enormous numbers. These are creatures like rats, fleas, lice, cockroaches, pigeons and sea-gulls, rarer in the prehuman world but now multiplying enormously. And, although they tend to be overlooked, we have also surrounded ourselves with less obvious creatures, like the teeming micro-organisms that have invaded the estuaries we have polluted, and the new mix of soil bacteria that is now found in the fields we have cultivated and altered with the addition of pesticides and fertilizers. These new combinations of micro-organisms provide new evolutionary opportunities. Pathogens with which we have previously been coexisting can be catapulted into new, short-term but terrible modes of existence. In short, they can become plagues.

We are not the only victims

Clumsy and genetically awkward as plague organisms are, they have a long evolutionary history. When plagues appear. they do so for a reason. And that reason has to do with the size of the host population and the degree to which it has disturbed its environment. Death, as somebody once said, is Nature's way of telling you to slow down.

We must be careful here, however. It is easy, and appears straightforward, to think of plagues as a form of population control. keeping our numbers in check when we have expanded too much for our resources. But this explanation, while it appears superficially to make sense, is redolent of popular views of ecology. Biologists hate this kind of thinking, because it implies that kindly but firm old Mother Evolution exhibits foresight, and that she somehow uses plagues deliberately to keep populations in balance.

Indeed, it is obvious that plagues have not evolved simply to control populations. After all, this would run counter to the central point of evolution, which is that successful organisms are successful because they can outbreed their competitors. Population control may be an effect of plagues, but it cannot be their *raison d'être*. The end result, however, is the same. When a population of any species – humans, rats, diatoms floating in the ocean – becomes too large for its resources, this provides new opportunities for a small subset of the many diseases that normally afflict it. Among these teeming pathogen populations, the ones that can cause plagues are normally rare. But when the population of hosts happens to explode in numbers, and to modify its environment in the process, the most virulent pathogens can suddenly gain a brief advantage. These dangerous pathogens are able to utilize this new brief ecological niche that is provided by the sudden superabundance of hosts. After the plague, the most virulent strains disappear or fall to low levels, and the host species and its pathogens soon revert to their previous uneasy balance.

If plagues really are on the increase in our species, this is telling us something very important about what we are doing to ourselves and our planet, something that we ignore at our peril. But ironically, it is the very plagues we fear so much that have helped to keep the world in balance for so long. And as the world's premier disturbers of ecosystems, we are helping to unleash diseases not only in our own species, but in others. We are justifiably worried about the appearance of new diseases that have emerged from the remnants of the natural world and that threaten us. But we tend to forget or ignore the fact that the flow is not one way. While we may not be contributing many diseases directly to other species, we are none the less causing plagues to spread among them that would not happen otherwise. These plagues threaten the survival of the world of nature at exactly the time at which it is most vulnerable.

This book began with the impact of a terrible disease on humans who have invaded and disturbed the rainforest. At the end of the book we will visit the rainforest again, and look at it with new eyes. The world harbours many forms of disease, and when we disturb its balance we are not their only victims.

2

The penumbra of disease

> The deviation of man from the state in which he was
> originally placed by nature seems to have proved to
> him a prolific source of diseases.
>
> Edward Jenner, 1798

The eyes of the hippopotamus

Loreto, six provinces that total a quarter of a million square kilometres of rainforest surrounding the city of Iquitos, accounts for about half of the lowland rainforest area of eastern Peru. Wilma Casanova de Caspía and her husband, Martín Caspía, are now in charge of epidemiology in the region. The two of them were kept very busy indeed during the plague year of 1991, for there were 27,000 cases of cholera in this region of about 600,000 people, with an official death toll of 426. The number of cases dropped to 6000 in 1992, and to 5500 in 1993. By July 1994, when I talked to her, there had only been 730 new cases, and no deaths. Cholera had been faced down, but hardly conquered.

Until the disaster of 1991, there had been no attempt to collect epidemiological information on this huge area. Now Wilma sends a weekly report to Lima on data collected from travelling doctors and nurses, who communicate with her through a network of radio stations. While there were, until 1994, only about fifteen medical professionals working in this vast area, the government recently doubled the salaries of health workers in the field from 400 to 800 dollars a month. As a result there are now about fifty doctors and nurses who report their activities to Wilma.

Most of the diseases that Wilma sees in the crowded shantytowns and in her numerous boat trips up and down the Amazon's tributaries are not plagues. Yet some of them can become plagues with astonishing rapidity, and others may become plagues in the future.

Plagues are the tip of the iceberg. Richard Guerrant of the University of Virginia has pointed out that since most cases of infectious disease occur in the tropics, it might be more appropriate to call them the eyes of the hippopotamus. The body of the hippo, hidden from view, is made up of the overwhelming majority of deaths due to infectious illnesses. These illnesses are not the result of what we call plagues. At the risk of mixing a metaphor, we can think of them as being caused by a kind of deadly shadow or penumbra of pathogens with which we are all surrounded. For example, a tiny fraction of the many deaths from diarrhoeal diseases are due to the terrible scourge of cholera. Most are the result of the unavoidable penumbra of pathogens that surrounds us, particularly those of us who live in the Third World.

Of course, even the pathogens that make up this penumbra threaten us only rarely – though the threats are far more numerous among the world's poverty-stricken tropical populations. Most of the time, these organisms simply cannot cause plagues among their hosts without jeopardizing their own survival. Instead, the major part of their evolution has been directed towards longer-term survival strategies. When pathogens step out of bounds they endanger themselves at least as much as the populations of human hosts in which they run amok. But most pathogens, most of the time, are no more and no less virulent than their long-term survival permits. And they tend, as we will see, to have evolved to become much cleverer and more subtle about their survival than their clumsy relatives that give rise to plagues.

Our defences are correspondingly ingenious. We have lived for a long time with the threat of these endemic diseases – far longer indeed than the relatively short span of geological time during which the species *Homo sapiens* has lived on the planet. The result has been a fluctuating but – in terms of our survival – bearable amount of sickness and mortality. It is when this delicate balance between life and death has been disturbed that plagues arise.

The risk from this disturbance is not spread equally over the planet. Diseases are more prevalent in the tropics. When we look in detail at how diseases work, we will find that this is so in large part because it is less demanding to be a tropical disease than it is to be a disease in the temperate zones. The sheer abundance of possible transmitting agents, along with the continual heat and humidity, mean that tropical disease organisms do not need to be as resourceful as those that must survive in cooler, more seasonal climes. Tropical disease organisms are hard to eradicate because of their sheer numbers and because of the many opportunities that the tropical environment provides for them, not because they are better at causing diseases than their temperate cousins.

The illnesses that Wilma deals with daily are the major part of the penumbra of diseases that surround us, and from which plagues can spring. They are startlingly different from those found in the First World, though the statistics she gathers would seem very familiar to an American doctor at the beginning of this century. The biggest killers are respiratory diseases like pneumonia and influenza. These peak during the long gloomy wet season when water finds its way into every hut through holes in its flimsy walls and roof, and the inhabitants are crowded together for long periods. These diseases' victims are chiefly babies, children under five, and very old people, and the numbers are frightful.

The omnipresent rainwater has a further effect, washing sewage into the wells on which the people of the shantytowns surrounding Iquitos depend. But during the wet season the drenching rains at least dilute the sewage, usually keeping bacterial contamination to a relatively safe level. As the water level drops during the dry season, the contamination in the shallow wells builds up. Now it is the turn of diarrhoeal diseases to sweep the communities, again chiefly targeting babies and young children. These illnesses are for the most part far less glamorous than cholera. They are caused by a grubby mixture of organisms: bacteria like *Shigella* and its very close relative the enteropathogenic *E. coli*, protozoa like *Cryptosporidium*, and viruses such as the omnipresent rotaviruses and the enteric adenoviruses.* The sheer number of these agents

---

* Bacteria are simple cells with structures and metabolisms very different from the cells making up our bodies, which means that they can often be killed by antibiotics that do not harm us. They can be crudely classified into bacilli, which are rod-shaped, cocci, which are round, and spirochetes, which are shaped like little spirals. Viruses, far simpler cannot multiply outside our cells, but unfortunately they spend most of their time hiding there, making them very difficult to attack. And protozoa like *Cryptosporidium* are single-celled organisms with metabolisms very like our own. The agents that can kill these protozoa and the even more elaborate parasitic worms must be used with great care, for they can often kill us as well.

makes them very difficult to diagnose, though the victims usually respond readily to clean water and rehydration therapy.

This deadly seesaw continues year after year. It is hard, Wilma told me, to see whether antibiotics and rehydration therapy have made any headway in the few years since their introduction. The effectiveness of data collection has grown so markedly in recent years that this increased reporting has probably masked any downward trend in the incidence of diseases. But there is no doubt that their real incidence is still higher than the best reported numbers. And it is highest in the shantytowns, where crowding and lack of sanitation continue to do their work as they have done in crowded and insanitary human habitations from time immemorial. Wilma has been able to report real progress against chickenpox, measles and whooping cough, however, since there are effective and cheap immunizations against these diseases. These are also the diseases that disproportionately affect the members of the Amazonian Indian tribes who have only recently come in contact with settlements.

In addition to such major concerns, Wilma tries to collect data on an astonishing variety of other diseases, many of them serious. Malaria is endemic – there are so many cases of vivax malaria that she cannot begin to keep track of them. More significant, because it can be so much more serious in its effects, is malignant falciparum malaria, of which there were 5000 cases reported in 1993. This disease used to be treatable with relatively inexpensive chloroquine, but the parasites are now becoming resistant. The one remaining effective drug, mefloquine (which I, as a privileged member of the First World, took religiously during and after my visit), is far too expensive for any but extreme emergency use.

Tuberculosis is the next big worry. Again, statistics are sketchy. During 1993, seven hundred cases of severe fulminant TB had been found among people who came into the hospitals of Loreto, and there were undoubtedly many less severe cases among people who did not seek medical help. Wilma's official estimate is that there are five life-threatening cases for every thousand people, but she told me that when she visits even the smallest village there are always one or two people who cough blood. A full treatment using the combination of antibiotics that can effect a cure costs about $50 and takes six months, and there is only enough money in the slender medical budget to treat at most two people in every thousand.

Leishmaniasis is common, particularly the virulent Amazonian kind that can sometimes eat away the flesh of the face. The disease, caused by a single-celled protozoan organism called a trypanosome, is transmitted by biting sandflies. The trypanosome actually hides within the cells of its victim, which makes it extremely difficult to treat. Although antimony compounds can effect a cure, a course of treatment costs $300 and is dangerous to the patient, so only the very severest cases receive treatment.

There are diseases in the hinterlands as well, including such horrors as fulminant hepatitis – often mistaken, Wilma told me, for yellow fever, which is also present, though relatively rare. Leprosy, which has been rare in Europe since medieval times, is still common in Loreto.

Together, these and other diseases form that penumbra of pathogens from which plagues can potentially spring, and today they are far more numerous and challenging than the diseases that afflict the populations living in the temperate zones. In Peru the official death rate among babies during their first year is one in ten (compared with one in a hundred in developed countries). But anecdotal evidence suggests that the rate may be more like three in ten among the Aymara and Quechua of Peru's southern highlands, and in the Amazon it is probably much higher than the reported rate as well. About twenty per cent of these deaths are due to enteric diseases, and a full third to diseases of the lungs – pneumonia, influenza and acute bronchitis. In the humid tropics these diseases are all in a day's work for a public health nurse. They form the usual drumbeat of endemic diseases that are found, to one degree or another, in Africa, southern Asia and tropical America.

In Peru there are so many diseases, and they are often so difficult to diagnose, that the fact that their cumulative death toll is far greater than the toll of the cholera epidemic passes almost unnoticed. It was the cholera epidemic, with its frightful and easily recognized symptoms, that shocked the world. And now, with the epidemic receding, Wilma's resources for fighting the endemic diseases are receding as well, as her nation's attention shifts elsewhere. Meanwhile she is doing her best to get on with her life, taking care of her new baby while trying to keep up with the endless but less dramatic public health crises that besiege this town on the edge of the world.

Crossing the ecosystem barrier

Peru straddles some of the highest peaks of the Andes just south of the equator, and is by many measures the most diverse country in the world. Its life zones range from the vast rainforests of the Amazon basin in the east, through dense and intensely green cloud forests that cover the rain-soaked eastern slopes of the Andes, to a bewildering maze of mini-ecosystems snaking among the high mountain valleys and the precipitous canyons of the western Andes. The coast itself ranges from the almost rainless deserts of the south to regions of startlingly wet tropical forests in the far north. The lowland rainforests of south-eastern Peru are home to the world's largest diversity of land animals and plants, with almost 1000 species of birds, 200 of mammals and 1200 species of butterflies. All this diversity has provided much that we find useful: Peru's varied ecosystems have given us tomatoes, potatoes, beans, peppers, squash, chocolate and peanuts.

The population, mixed now from migrations within the region and from the European invasion, is made up of about fifty per cent pure-blooded Indians, 37 per cent mestizos (mixed blood), and the remainder of primarily European origin. They include highland Quechua and Aymara Indians, who have become physiologically adapted to life at an altitude of 4300 metres, the very edge of what is possible for humans, and who have immense capacities to do physical work under these conditions. And they also include lowland Indian tribes, still living in the Stone Age, who have yet to be contacted by civilization.* The peoples are as varied as the diseases that afflict them.

A glimpse of the true complexity of the interactions between people and disease in Peru was given to me by Dr Juan Murillo of the Institute for Tropical Medicine at Lima's San Marcos National University.

First of all, as he pointed out, Peru is a country in upheaval. When I visited the campus of San Marcos, the presence of the army was not as obvious as it had been in the months following the arrest of Abimael

---

* The interactions among these various worlds can still be violent. The week before I arrived in the rainforest of Manù Park in south-eastern Peru, my guide had been shot at from the river bank by a member of one of these uncontacted tribes. Luckily, the arrow broke and bounced harmlessly off the side of the dugout canoe, leaving him in possession of a feathered shaft that was unlike anything the local inhabitants had seen.

Guzman, the leader of the guerrilla group known as Shining Path, in September 1992. After his capture several bombs had exploded on the campus, part of the Shining Path's desperate attempts to rally its followers after the blow to its leadership. Ironically, the Shining Path has helped to pressure the Fujimori government into instituting drastic economic reforms that show some sign of working. Nonetheless, population growth is still unrelenting, poverty is widespread, and these reforms have, in the short term, increased economic pressures, particularly in the highlands. This is because government-run mining and agricultural operations are rapidly being privatized, and thousands of workers in the rural areas are being laid off.

All these economic dislocations, along with the activities of the Shining Path, have resulted in an acceleration of the long-continued massive movement of highland Indians into both the cities of the coast and the settlements of the lowland forests to the east. This movement has spread diseases that were formerly confined to fairly small regions, infections like histolytic amoebiasis, the vicious haemolytic disease *verruga peruana*, and Chagas' disease. Although the *verruga* epidemic receded during the heyday of DDT spraying from 1959 to 1974, it is now coming back in a milder form as the sandflies that carry it multiply once again.

These migrations have also brought the highland peoples into contact with tropical diseases for the first time. The most dramatic of these collisions was a severe epidemic of vivax malaria among highland Indians in the 1930s. This malaria, relatively mild under most circumstances, was extremely virulent in its effects among the unexposed Indians. Now, with vivax widespread throughout most of the country, it has again become much less severe, causing a set of symptoms more like those that it exhibits elsewhere in the world.

The cholera bacillus, Murillo is convinced, has been present in Peru in small numbers for decades. He first cultured it from a patient in 1983, eight years before the outbreak. With help from researchers from Johns Hopkins University, from 1985 onward he repeatedly found strains of cholera bacilli in people suffering from severe diarrhoea. These strains were different from the deadly strain that caused the outbreak, but they still manufactured small quantities of the toxic protein that triggers the disease. They seem to have been overwhelmed by the introduction of the new strain from southern Asia in 1991 that triggered the epidemic. Yet the potential for an outbreak seems to have been present in the population even before the arrival of the Asian strain.

The situation facing nurse Wilma, Juan Murillo, and their colleagues is not new, and until recently it was not confined to the tropics. People who live in the relatively disease-free countries of northern Europe and North America do not have to go back many generations before they discover that their ancestors faced a collection of diseases that, while not quite so diverse, was equally threatening.

We are not the only species to be afflicted by a penumbra of diseases, although of course we make the most fuss about it. A similar penumbra surrounds most other animals and plants. While many of our diseases are unique to us, we share some of our pathogens with other animals – and, remarkably, even in a few cases with plants. It is more than a coincidence that so many of our drugs are derived from the plant world, for plants are attacked by pathogens too.

3

The worst of times

Whereas we have found, that of 100 quick Conception
about 36 of them die before they be six years old, and
that perhaps but one surviveth 76. we having seven
decads between 6 and 76, we sought six mean pro
portional numbers between 64, the remainder, living
at six years, and the one, which survives 76 . . .

John Graunt, 166?

What did our ancestors die of, and in what numbers? Occasionally we
can get a vivid glimpse of an earlier time when mortality was appalling
and the causes of death were very different from those faced by any
population in the world today.

The draper John Graunt lived in London during the tumultuous
years of England's Civil War and the grim dictatorship of Oliver Crom-
well that followed it. He became fascinated by London's Bills of Mortality, which appeared weekly and which were consulted like traffic
reports by the privileged who used them to calculate when they should
flee the city.

In pulling the numbers together and looking for patterns in them,
Graunt became the world's first demographer. He examined 22 of the
years between 1629 and 1660. In his day an autopsy or coroner's report
was a rarity. Instead, the information on cause of death was gathered
by a sworn company of searchers, made up for the most part of 'antient
matrons', who arrived at the place of death, made their determination,
and (depending on whether or not they were literate) wrote down the
particulars or else memorized them and told them to the parish clerk.

Though the searchers knew little of medicine, by long familiarity

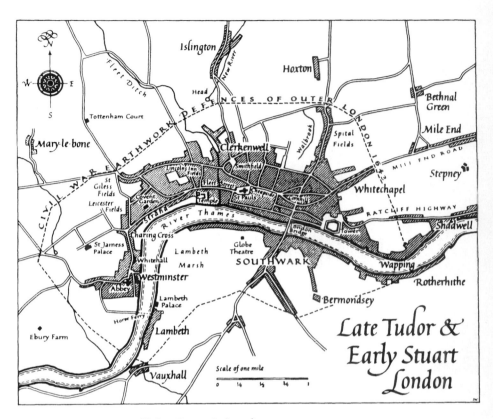

FIG. 3-1 A map of John Graunt's London

with their parishes they could distinguish a hundred or more causes of death. In two of the 22 years examined by Graunt the greatest cause was plague, but in most years plague was outnumbered by far by deaths due to 'consumption and cough', which accounted for a fifth of the 22-year total of almost a quarter of a million deaths. Stillbirths and infant mortality accounted for another tenth, and this number, as Graunt himself realized, was undoubtedly a woeful underestimate. I have counted up the number that died of a hideous collection of diseases caused by infections, ranging from 'teeth and worms' through 'purples and spotted fever' to the King's Evil (scrofula, a form of tuberculosis), and find that it makes up 66 per cent of the total (again undoubtedly an underestimate, for much infant mortality due to infectious disease was certainly not included). Old age, on the other hand, accounted for an encouraging seven per cent, showing that a small fraction of the population could still manage to make it through these frightful perils.

Oddities abound in the tables. Cancer is a rarity. There are a mere two deaths attributed to excessive drinking, which undoubtedly reflects the unavailability of anything stronger than fortified wine. And there are a mere six deaths attributed to leprosy, though two hundred years earlier this number would have been much higher. While there is much apoplexy, accounting for a tenth of the total, there is no sign, in these mortality tabulations or any others until the last part of the nineteenth century, of anything resembling heart failure – except perhaps for a small number who are recorded as having died of fright.

Of the three great killers in the Western world today, heart disease, cancer and stroke, only the last is likely to have contributed substantial numbers to Graunt's tables, presumably as some subset of those who died of apoplexy. The world has changed immensely since Graunt's time.

Each Bill of Mortality, of course, gave only a snapshot of that week's deaths, lumping together people who died of old age with stillborn infants and everybody in between. Graunt was the first to go a step further, by constructing the first *life table*. In the process he helped to found the science of statistics and the life insurance industry. Unfortunately, he was not sure of the ages at which people died, so he had to make some guesses. He could be reasonably confident about those who had died in infancy and early childhood, and about those who had died of old age, but he was forced to interpolate the intervening numbers. Starting with 100 inhabitants of the city, he estimated how many would still be alive at age 6, age 16, 26 and so on. Figure 3-2 shows the result.

Infant mortality was probably higher than the figures obtained by Graunt, which showed that about a third of the population had died by the age of six. What is particularly striking about Graunt's table is that if his guesses were right this terrible mortality must have continued unabated throughout adulthood – only a quarter of the population remained alive by age 26 and only six out of every hundred by age 56. It is this continuing mortality that so strikingly illustrates the impact of then-incurable infectious diseases.

Graunt's curve is too smooth, of course. Indeed, we can construct a theoretical curve that makes an excellent fit to the data if we simply

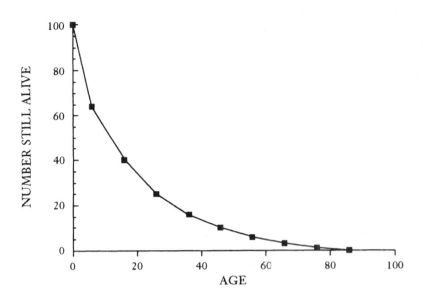

FIG. 3-2 John Graunt's mortality estimates for seventeenth-century London

assume that about 43 per cent of those remaining from the original 100 die during each succeeding decade. Mortality in the middle years was probably not as fierce as he presented it. But without precise information on the age of death it is impossible to determine how much the real curve deviated from Graunt's.

The pattern of mortality has changed in complex ways over time, and we do not fully understand why. There may have been more appalling periods in the history of human mortality than that of John Graunt, but we do not know of them. What fragmentary data we have suggest that things were somewhat better earlier, and we have good data showing that mortality eased after Graunt's time.

The earlier data are necessarily fragmentary and doubtful. It is possible for archaeologists today to construct life tables from skeletons at burial sites, where the age of death can be approximately inferred, and Figure 3–3 shows one from Catal Huyuk in what is now south-central Turkey, one of the earliest sites at which extensive agriculture was practised.

The people of Catal Huyuk suffered appalling infant mortality and none of them made it into what we would call old age. But at intermediate ages they may have done a bit better than the inhabitants of Restor-

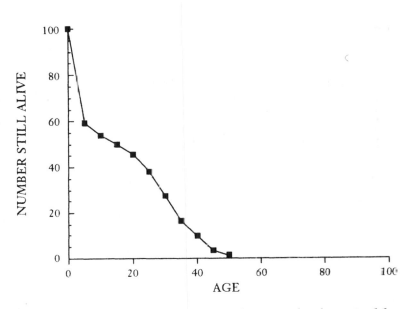

F I G. 3-3 Mortality data from Catal Huyuk in Anatolia, determined from burials that took place between 6500 and 5300 B C

ation London. Perhaps this is because the impact of infectious diseases on these early agriculturalists was smaller than the impact of environmental severity such as drought and starvation. These rigours would have disproportionately killed the very young and the very old, the most susceptible members of the population.

Certainly, food and resources were rather more plentiful for the inhabitants of Restoration London than they were for those of Catal Huyuk. Provided they could thread their way through the perils of all those infectious diseases, a small percentage of them could reach the calm waters of an old age not very different from our own. But the perils due to disease that lay in wait during these Londoners' middle years may have been even greater than those at Catal Huyuk 8000 years earlier.

Things are very different now. But, surprisingly, the change that has taken place since Graunt's time did not happen all at once following the discovery of microbes and the rise of official public health programmes in the nineteenth century. It has been a far more gradual process, and it began long before the work of Louis Pasteur and Robert Koch and the acceptance of the germ theory of disease.

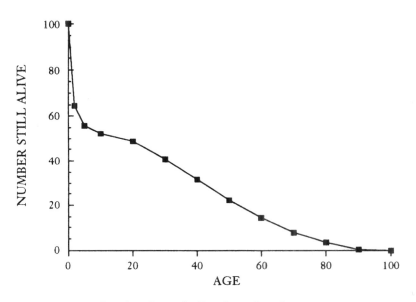

FIG. 3-4 Mortality data from the London of 1728

The next two graphs show this gradual improvement in mortality rates. In Figures 3-4 and 3-5 I have calculated life expectancies drawn from London's Bills of Mortality (compiled by John Marshall in 1831) for the years 1728 and 1830. Here, since the age at death was recorded, the data are far more precise. The 1728 graph (Figure 3-4) is strikingly different in shape from Graunt's, with immense infant mortality followed by a much slower decline in numbers during intermediate ages.

The 1728 curve looks very like that seen in Catal Huyuk 8000 years earlier, although a few people lived to a much greater age.

Had Graunt been able to calculate an equally accurate curve, it probably would have looked very similar, though the 'bulge' in the middle would have been smaller. We suspect this because a century later, in 1830, the 'bulge' has grown more pronounced and the first year or two of life a little less lethal (Figure 3-5).

The next figure (Figure 3-6) shows a curve I have calculated from a life table for the English and Welsh population of 1861.

You can see that the trend towards reduced infant mortality and enhanced survival in the middle years, a trend that had begun after Graunt's time, was accelerating. There was still substantial infant mortality in the England of 1861, but once that hurdle was passed there

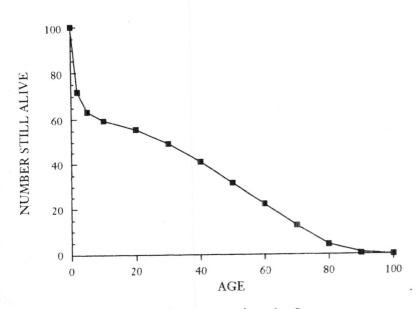

was a reasonable chance of living at least until late middle age. And this shift had begun to happen before public health became a priority, and even before the germ theory of disease. It was taking place before the 1860s, when Parliament passed the first bills to improve the condition of England's water supplies.

The change during the next century was by far the most dramatic that any human population has undergone. The true impact of our conquest of infectious disease is seen in Figure 3-7, in which the fate of 100 modern Americans is followed over time. This curve is typical of the populations of industrialized societies.

At the present time, infant mortality is a tiny dip in the graph, and substantial mortality does not begin to take place until middle age. (The striking difference between the races is largely an economic rather than a racial difference, by the way; the subgroup of whites who have the same income level as blacks show the same mortality curve. And mortality due to AIDS is too small to have had a significant impact on the shape of the curve.)

I know of no better way to illustrate the change in our lives resulting from the near-conquest of infectious disease. But the graphs illustrate that the effects on lifespan of this conquest, while they have certainly

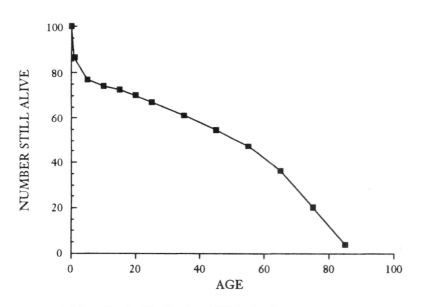

FIG. 3-6 Mortality for England and Wales in the year 1861

accelerated lately, were more gradual than is generally supposed. What happened? Why were plagues releasing their grip on us even before we realized what caused them? And what does this tell us about the nature of the plagues themselves? Could their grip tighten again, and if so why?

Bacterial Uriah Heeps

> There is nothing to be gained, in an evolutionary sense, by the capacity to cause illness and death. Pathogenicity may be something more frightening to them than to us.
>
> Lewis Thomas, *Notes of a Biology-Watcher: Germs* (1972)

In a recent book, *Evolution of Infectious Disease*, Paul Ewald argues that the late Lewis Thomas (along with many other people who have tried to understand epidemic diseases over the last century and a half) was wrong. There is nothing about the evolutionary process, Ewald points out, that dictates that diseases and hosts should learn to get along with each other. Disease organisms do not necessarily become more benign

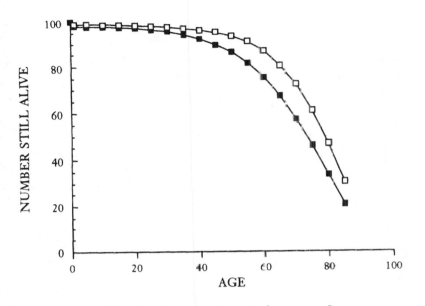

FIG. 3-7 Mortality of modern Americans in the years 1985–7

over time, finally reaching a kind of bland *modus vivendi* with their hosts, unless they happen to be selected to do so. But they do, he suggests, often become more benign if it is hard for them to transfer from one host to another. For instance, if the water supply in an area of human habitation has been cleaned up, vicious bacteria that used to cause severe enteric diseases like cholera and typhoid will tend to be replaced by more benign variants of the same bacteria.

Such replacement is not confined to bacteria. An AIDS virus that is vicious, fast-acting and easily transmitted in East Africa, where societal conditions allow it to be spread rapidly, is found at much lower levels in West Africa. There, where conditions seem to be less favourable to viral spread, another closely-related but less virulent virus predominates.

It is fairly easy to understand the evolutionary reason for this lessening of virulence. Nasty pathogens that cannot get from host to host easily are doomed, for they may kill their host before they have a chance to spread. But we can turn this argument on its head. Why is it that when things are easy for the parasites – a multitude of hosts, easy transfer from one host to another – they often tend to be more ferocious? Why, in short, do parasites tend to act like tiny Uriah Heeps? Here, you will recall, was the memorable character in Dickens's *David Copperfield* who was sickeningly ingratiating when he was in a subordinate position, but who became vicious and overbearing when he had acquired (as he thought) a little power.

Occasionally, for reasons that we are slowly beginning to comprehend, disease organisms throw off their subservient guise and cause the great disasters of plagues. But it is puzzling that they should do this, simply because there happen to be many hosts, leading to favourable conditions for their spread. Why should they not continue their mode of peaceful coexistence, simply in greater numbers? When more virulent strains appear they are, after all, putting themselves at risk, since such strains reduce the survival of their hosts.

If we can distance ourselves from the dreadful facts of these plagues long enough to put them in an evolutionary context, it is possible to draw two conclusions. The first arises from the word *occasionally*. Plagues are, on the whole, rare – and it is lucky for our species that they are. At this very moment, our guts and those of our domestic animals are filled with the innocuous relatives of many different plague organisms, and others are found in the soil and elsewhere. They spend most of their time minding their own business, dining off what they can get, and keeping a commendably low profile. Most of these parasites trouble us very little and go to uncounted and unrecorded graves by the trillion. Uriah Heep, too, spent years in his subordinate position, rubbing his hands together and assuring everyone how 'umble he was. He might have gone to his grave 'umble to the last, had his plans never come to fruition.

Plagues are *sporadic*, usually brief, and often associated with unusual circumstances. Like Uriah Heep, they cannot strike anywhere at any time, but must in effect bide their opportunity.

The second conclusion is that there must be an evolutionary reason for the increased virulence of plague organisms. Ultimately, that reason must be their survival. Lewis Thomas was wrong when he said that parasites had nothing to gain in an evolutionary sense by causing illness and death. Parasites have everything to gain if the only way in which they can spread is by causing such havoc. But I think Thomas was right when he said, with some anthropomorphic hyperbole, that pathogenicity may be more frightening to them than to us. They embark on a dangerous course when they drop their disguise. When Uriah Heep revealed his true character, at the end of Dickens's novel, he also revealed the fragility of his vicious scheme and was quickly brought down. And when parasites, like Uriah Heep, embark on the high-profile course of disease and death for their hosts, they do so at their own peril. They put not only their hosts, but themselves in harm's way. The imperative that forces them to this extremity is the necessity of ensuring their own immediate survival, nothing else. And although increased virulence may ensure their survival over the short term, the conditions that engender plagues always end sooner or later. This always, *always* means that plague pathogens will not survive in the long term unless some of them manage subsequently to lose that virulence and revert to the habits of their low-profile relatives.

Yet the nagging question persists. Why do plagues happen at all? Why don't all pathogens keep a low profile all the time? The question can be answered in two ways, one trivial, the other deep and subtle. One trivial but true answer is that, when pathogens are presented with a short-term opportunity, variant strains that can exploit it will always arise or will increase from formerly low numbers – even at the risk of putting themselves in danger later on. Like a bank robber concerned only with immediate gain, they take advantage of the new niche, no matter how fleeting and risky it might be, as soon as it appears. The result can be a plague. Another possibility is that the virulence of a plague is an accident – perhaps a product of the sheer numbers of plague organisms in the host.

The deep and subtle answer is true as well, but it has only begun to be appreciated recently by scientists as our knowledge of the molecular biology of plague organisms has grown. It has to do with the evolutionary history of plagues. This history stretches far back, to one before our own species appeared on the planet. Again and again, all during that huge span of time, conditions arose repeatedly that allowed plagues to spread through the populations of those remote ancestors of ours.

How do we know this? In some cases, the traces of those early plagues are written in our genes. In others, they are written in the genes of the plague organisms themselves. Fleeting as those 'plague niches' were in both time and space, they have turned up so often through the ages that pathogens have by now become experts at exploiting them.

There has been and continues to be much genetic borrowing among pathogens. It is now possible for us to detect traces of this borrowing, particularly among bacteria, because the toxin genes and other virulence genes that are carried by many bacteria reveal their history in their DNA. It is becoming clear that these genes have done most of their evolving, not in the bacteria in which we happen to find them, but somewhere else. Particularly in the microscopic world of bacteria and viruses, evolution can be a great borrower – if a gene evolves that happens to do a particular job very well, it is not long before it will spread widely among many other, sometimes quite distantly related, pathogens.

In acquiring these genes, our pathogens have been handed tools that they are able to make deadly use of in emergency short-term situations. But why should these tools have evolved in the first place? It is as if assault weapons were to grow on the trees of Central Park, available for anyone to pluck down. The tools themselves have evolved because plagues occasionally afflict every species, forming an extreme but essential part of evolution.

Thus, when the niche presents itself again, the pathogens are immediately able to exploit it. It may require only a mutation in a single gene, or the acquisition of a piece of DNA borrowed from another micro-organism. Plagues can happen because the organisms that cause them are only a few steps away, genetically speaking, from their less harmful relatives that inhabit all of us all the time. And they stop happening because the conditions that caused them quickly dissipate, leaving the mutant pathogens without an ecological niche. They are as vulnerable as bank robbers whose getaway car has driven away without them. Yet, because they have been shaped by this long evolutionary history, the less virulent pathogens from which they arise are always ready to strike again.

Scientists began to exploit the vulnerability of plague organisms a century ago, rapidly winning battles against typhoid, typhus, bubonic plague, tuberculosis, malaria, and many other killers. Yet in other cases, astonishingly, the plagues disappeared by themselves and we still do not know why. Bubonic plague, which had afflicted the population of London for centuries on a regular basis, suddenly disappeared after the great plague of 1665–6. What happened? Certainly it was not a conscious effort at eradication – the medicine that was practised at the time was still primarily concerned with balancing the four humours of the body, and nothing was known of the existence of bacteria. We do not know what it was that the inhabitants of London and the other large cities of Europe did differently that brought about the cessation of plague, though we will shortly join many others in speculating about it.

Malaria, too, suddenly disappeared from northern Europe, although this time the disappearance took place much later, halfway through the nineteenth century. Even so, this was still fifty years before the connection was made between malaria and mosquitoes. We have a slightly better idea of why this event happened, and it is a story that will be told in its place. But again it is obvious that there was no conscious attempt to eradicate malaria on the part of the people who benefited from the disease's disappearance.

It is possible that the plagues we are faced with today – AIDS, resurgent tuberculosis, the enteric diseases that cause so much havoc in the Third World – will turn out to be similarly, unexpectedly vulnerable, if we can determine when and where to strike at them.

The story of each plague can be seen as a variation on this evolutionary theme, a dance between survival and extinction in which both host and parasite take part. Nowhere is this tale more vividly illustrated than in the story of the eponymous plague itself, the Black Death.

PART TWO

*Chief monster that hast
plagued the nations yet . . .*

4

Four tales from the New Decameron

Between March and July of the year in question ... it is reliably thought that over a hundred thousand human lives were extinguished within the walls of the city of Florence[.] Yet before this lethal catastrophe fell upon the city, it is doubtful whether anyone would have guessed it contained so many inhabitants.

Giovanni Boccaccio, *The Decameron* (tr. G. H. McWilliam)

In the summer of 1986, a Canadian government helicopter was carrying out a routine aerial survey that included some flights over remote Banks Island. This island, covered with low tundra vegetation, lies in the Northwest Territories, far above the Arctic Circle. In previous years, dead and dying muskoxen had often been spotted by aerial surveys. But now, as the helicopter flew over the peaceful landscape, the pilot suddenly saw the bodies of almost sixty wild muskoxen, scattered over several square miles. Had they been wantonly shot by airborne hunters? Veterinarians who were helicoptered to the site found that the truth was even more unnerving. All were discovered to have died, within the space of a few days, from acute respiratory infections.

During that summer, 122 sick and dead animals were found over an area of 1000 square kilometres. Muskoxen, which were driven to local extinction on the island a century ago by overhunting, have rebounded to a population size of 25,000, a size that has remained fairly stable. Before these dead and dying animals were examined, it had been assumed that shortage of food was now limiting the size of this recovered population.

Microscopic investigation of various tissues showed that their deaths were caused by the bacillus *Yersinia pseudotuberculosis*. This organism is a relative of the far more infamous bacterium *Y. pestis* which is responsible for human bubonic plague. The animals were riddled with infection, which had been as devastating to this herd of muskoxen as the great plagues of Europe have been to our own species. And this was not an isolated incident. Occasional fatal outbreaks of *Yersinia* have been recorded among different species of animals in many parts of North America, Europe and Asia, and it seems certain that innumerable similar incidents have gone unrecorded.

Most of us probably assume that such plagues are visited on the human species alone. Of course this is not true. Plagues are an important part of the natural world, though most of them are easy to ignore because they have no direct effect on us. Each kind of plague can be seen as an evolutionary detective story, complete with a vast assortment of clues. In a classic detective story the criminal has some weakness or failing that can eventually lead to his or her downfall. The same is true of the organisms that cause plagues, and the trick is to find it, though sometimes the clues are buried very deeply.

No trail of clues is more devious or fascinating than the one that has led to our current understanding of bubonic plague. This horrendous disease has, by one estimate, been responsible for as many as 200 million human deaths during our recorded history. To comprehend the plague in all its aspects, and to probe for any weaknesses that it might have, we must tell four stories. The first shows how it has affected our own species, the second recounts its effect on other species of animals, the third is the tale of the plague bacillus itself, and the fourth is an account of the precarious interaction of the bacillus with fleas, showing the desperate lengths to which the bacillus must go to perpetuate itself. By the end of this chapter and its four stories you will be able to make connections between the events that happened to our own species in previous centuries (and are still happening to us today in places as different as New Mexico, Tanzania and India) and the disease that struck down those innocent muskoxen on the tundra a few years ago.

The Human Tale

In the year 542, the Byzantine Empire was in its usual state of upheaval, pursuing war on many fronts. General Belisarius, the head of its armies, was fighting an indecisive war in Persia. The campaign was something of a distraction for him, sandwiched between two much more successful campaigns against the fierce armies that occupied the remains of the Western Roman Empire. He was labouring to accomplish the simple but sweeping goal of his master, the Emperor Justinian, which was nothing less than to re-establish the Roman Empire in all its glory.

Suddenly, in the midst of Belisarius' attempt to execute Justinian's grand design, a disaster overtook Byzantium. A plague of huge proportions swept out of Egypt, carved its way through the eastern Mediterranean, and went on to devastate the rest of the known world.

The historian Procopius was an eyewitness to the effects of the pestilence when it arrived in Constantinople. Groping for an explanation, he could only attribute it to God, for 'it embraced the entire world, and blighted the lives of all men, though differing from each other in the most marked degree, respecting neither sex nor age.' He thought that the plague could not have been transmitted by anything other than the capricious will of the Deity, for he observed that many nurses and physicians were unaffected even though they were working in the midst of those who had been felled.

Procopius described the symptoms of the pestilence vividly, so we can be pretty certain that it was in fact bubonic plague. He recounted the panic that the plague caused among superstitious people – many of the victims, he reported, claimed that they had been visited by demons even before the disease's onset. This might have been due to the general hysteria of the time, or simply have been hallucinations brought on by the fever as it developed. But there were no mystical trappings to Procopius' account of the course of the disease itself, which is clinically accurate – the great swellings called buboes that appeared in the groin and the armpits as the lymph nodes became engorged, the delirium and frantic restlessness suffered by some victims and the comatose state of others. He observed, accurately, that if the buboes broke open there was a chance the victims would recover, though they were sometimes crippled. Otherwise death almost certainly ensued within a few days.

The plague raged in the city for three months, and so numerous were the corpses that many of them were simply cast into the hollow towers of some uncompleted fortifications in nearby Galata, across the Golden Horn. Procopius calculated that as many as 300,000 inhabitants died, which may be an exaggeration since the total population of the city has been estimated by modern historians to be smaller than that. But, as Boccaccio later observed during the time of the Black Death, it was easier to count corpses than it was to count the living poor who infested the ancient alleyways of crowded medieval cities. Procopius' estimate may have been right.

Whatever the exact numbers, there is no doubt that, as with other outbreaks of bubonic plague, the mortality was very high. Yet in spite of the disaster, Justinian (who according to Procopius was himself briefly ill with the plague but recovered) immediately took advantage of the devastation in nearby regions to invade Armenia. So powerful were the military machines of Byzantium and the states that surrounded it that warfare continued unabated both during and after the ravages of the plague.

Yet this and previous pestilences had helped to change the balance of history. All of Europe was affected by the Plague of Justinian – five years after it first appeared in Constantinople, the disease had swept west three thousand kilometres to Ireland. While other, less well-documented epidemics had preceded this pandemic during the decline of the Roman Empire, they might have been anthrax or typhus rather than plague. There have been many speculations that Justinian's Plague was the event that precipitated the final collapse of the tattered remnants of the Western Roman Empire and the advent of the Dark Ages.

Without this disaster, history might have gone in quite a different direction. Before the plague struck, Belisarius and others of Justinian's generals had won a series of victories, both in North Africa and Italy to the west and in the countries of the Levant. The growing Byzantine Empire had been strong, healthy and well-populated – perhaps, just perhaps, if the plague had not taken place Justinian might indeed have re-established the elusive glory of Rome.

The plague continued to be an important factor in frustrating the plans of Justinian and his successors, for not only did it sweep through his realm in the three years following 542, but it recurred at intervals, sometimes no more than three or four years apart, for decades thereafter. Its malign effects extended well into the seventh century. And the longer the intervals between subsequent epidemics, the greater the mortality among young people, since many unexposed children were born in the lulls between the outbreaks.

These repeated epidemics knocked the population down faster than it could recover. The historian Josiah Russell has estimated that the proportional impact on the population of Europe may have been even greater than it was during the later Black Death. The population of western Europe may have been reduced by as much as fifty per cent by the year 600.

The periodicity of the outbreaks, which was also seen in the Black Death some eight hundred years later, is now suspected to mirror the rise and fall of populations of rats and other rodents in the affected areas. As rats increased in number, the plague bacillus would begin to spread among them, so that both the rats and the people who lived in close association with them would be decimated. If the rat population fell dramatically, it might be a decade or more before it could recover, but when it did the plague would reappear. If the rats were not as severely affected, their population could rebound in three or four years. All this had, it might be supposed, relatively little to do with the humans involved, who seem merely to have been hapless victims caught on the perimeter of this rodent–bacillus population cycle. But is that really true?

We must begin by asking where Justinian's Plague came from, and why it appeared when it did. There is general agreement, among observers at the time as well as present-day historians, that the much later Black Death of the fourteenth century originated in central Asia. We can never be sure, but there is also a growing feeling among historians that the earlier plague of Justinian's time had its origin elsewhere. Procopius claimed that it originated in Egypt. Arab physicians of the time, who had themselves been confronted with some of the repeated outbreaks of plague that soon followed its first appearance in 542, thought that the disease had originated further south, in the uncharted regions of Abyssinia. Indeed, fragmentary historical accounts suggest that the plague has been endemic for many centuries in that remote country, where it was called *jaghalah*. An early Arabic medical compendium, published in the year 850, placed the origin further to the west, in the Sudan, where the Nile would have provided an easy route for the disease to spread into Egypt.

None of these early accounts implicates Asia. Indeed, there is even some evidence that plague was rare or unknown in central Asia at the time. After the death of Procopius, the Byzantine historian Theophylact Simocattes continued his mentor's task of recording the history of the Eastern Empire. Simocattes told of Turkish potentates from central Asia, visiting the Byzantine court soon after Justinian's time, who boasted that the plague and other contagious diseases were quite unknown in their part of the world.

So, an African rather than an Asian origin of the Plague of Justinian is a very real possibility. At the present time there are known to be foci of plague among the rodents of eastern Africa, and outbreaks of the disease have been traced unmistakably to that area since the sixteenth century. Possibly the plague that lurks there now could have been a recent introduction to Africa, from other known foci in Asia perhaps, at some time subsequent to the sixth century. But this seems unlikely, for so many different rodents native to East Africa are currently involved, and the situation is so complicated, that it seems certain that the plague bacillus and its many rodent hosts have had a long and complex evolutionary history in Africa.

The records of such a distant time are never unequivocal, however. Asia cannot be ruled out as a source of Justinian's Plague. Large populations of rodents have swarmed in central Asia since long before humans first moved into the region a million or more years ago. Some of these rodents came in close contact with humans, and were hunted for food and even for their skins. Marco Polo, relaying tales of the region of Tartary that lay to the north of China, says: '. . . there are various small beasts of the marten or weasel kind, and those which bear the name of Pharaoh's mice. The swarms of the latter are incredible, but the Tartars employ such ingenious devices for catching them, that none can escape their hands.' The term 'Pharaoh's mice' had originally been applied to the mongooses of Egypt, so it is a little unclear which rodents Marco Polo meant, but the little animals he had seen in such numbers were probably tarbagans or Manchurian marmots, about the size of a squirrel. They have been shown by Russian microbiologists to harbour the plague bacillus, and yet Marco Polo does not mention disease among the Tartars.

Just as in Africa, the rodent population of the steppes of central Asia is a diverse one, and there is an astounding number of different rodents living there that carry many different strains of the plague bacillus and of allied bacterial strains. In central Asia, as in Africa, plague has had a long and complex evolutionary history. The disease certainly goes back to a time long before there were any humans in the region.

In spite of the prevalence of plague among the wild rodents, we have only fragmentary evidence for early outbreaks of bubonic plague affecting the human populations of Asia. Chinese historians kept careful records, starting in 244 BC, of plagues and pestilences that devastated the country repeatedly. The numbers of such events are astonishing – from AD 37 to 1718 there were 234 recorded outbreaks. Chroniclers recorded one every three years on the average during the sixteenth and seventeenth centuries, when records were probably quite complete. They noted the provinces affected, and gave some account of the numbers of people who perished, but unfortunately they almost never made note of the plagues' symptoms – many of them could have been, and probably were, typhoid, smallpox or something else. Two accounts, however, do mention 'malignant buboes' and other symptoms that sound suspiciously like plague, one dating from AD 610 and the other from AD 642. If the plague was really that old in China, then it might by whatever devious route have been the source of Justinian's Plague. As we saw earlier, equally equivocal references to plague in India can be traced back, through the chronicle of the *Bhagvata Purana*, to as long ago as 1500 BC, suggesting that plague may be much older than the earliest Chinese records.

Whatever its origins, one thing seems plain. When we disturb its various ancient haunts, the plague bacillus comes forth. Human disturbances of this status quo in the ancient world were, like those of modern times, primarily due to trade and the movement of peoples. In the days of Justinian there was little contact with Asia, but Arab traders were beginning to penetrate Abyssinia and the even more remote regions further south in search of gold, slaves and ivory. They may have found more than they bargained for.

Yet in the centuries that followed Justinian's Plague, bubonic plague essentially vanished from the West. Many other dreadful diseases afflicted the inhabitants of Europe as they shuffled blindly through the very nadir of the Dark Ages. But bubonic plague died out in spite of what would appear to be ideal conditions for its persistence. Did this have something to do with the fact that the plague had so dramatically reduced the European population?

It is almost certainly not coincidental that eight centuries later, when bubonic plague re-emerged in what would later be known as the Black Death, this new scourge happened just when the population was beginning to increase and the outside world was again starting to impinge on the isolated European peninsula. Traders were venturing in substantial numbers along the Silk Road that stretched across Turkestan and central Asia to the empire of the Great Khan. They were lured not only by silk but by spices, and by the astounding tales of wonders brought back by Marco Polo. Like a monster in a horror film that seems to lie around every corner, the plague waited for them there as well.

Oddly, it appears that it was refugees from the Byzantine Empire who may have been the first to fall victim to this second great plague pandemic. In the fifth century, shortly before the Plague of Justinian, there had been a considerable religious upheaval (one of many) in Byzantium. The followers of Nestorius, bishop of Constantinople, were driven from the main body of the Orthodox Church because of their refusal to acknowledge Mary as the mother of God. Fiercely persecuted, they established settlements in the remotest parts of Asia, and some of them eventually came under the protection of Kublai Khan. As a result of this forced diaspora, communities of Nestorian Christians have persisted on the remote edges of the known world for centuries.

In the year 1338, an outbreak of a devastating disease killed many members of one of these communities, which was located on the shores of Lake Issyk-Kul not far from what is now the Kazakh–Chinese border. These deaths were quite unremarked at the time, and their cause was only discovered early in this century by the Russian archaeologist Daniel Abramovich Chwolson, who translated some headstones and found that the victims had probably died of bubonic plague.

This event has usually been cited as the beginning of the second great plague pandemic, which later became known as the Black Death. It may or may not have been, of course. The plague may have originated further to the east in China – or perhaps even in Africa. The trail of evidence is tenuous to say the least, and has grown cold in the intervening 650 years. Whatever its origin, and whatever the trigger may have been for its new spread, the plague found fertile ground.

At the beginning of the twelfth century, throughout Europe and the Mediterranean, populations had begun to grow after the dreadful stasis of the Dark Ages. During the 'little Renaissance' of that century political stability grew, new crops were introduced, and a middle class made its first tentative appearance. Agriculture began to improve – not much, but enough to make a difference. One of the reasons Europe had remained mired in the Dark Ages for so long was the pitiful yield of crops – on average, one seed had to be planted for every three obtained, which might seem to make agriculture an almost pointless exercise. Such dismal harvests actually persisted in Russia well into the nineteenth century, and helped foment the political discontent that eventually led to revolution. But elsewhere in Europe things began to get better as windmills, water wheels, horseshoes, the horse collar, and the mouldboard plough came into widespread use during and after the twelfth century. These inventions permitted land that had been allowed to revert to forest or marsh during the Dark Ages to be cleared and drained again.

Trade employing money rather than barter became widespread for the first time since the fall of the Roman Empire, and with this new prosperity towns began to grow. Both the towns and the countryside became surprisingly densely populated – Leopold Genicot writing in the Cambridge Economic History has estimated that many parts of rural France had populations that would not be surpassed until the beginning of the twentieth century. German towns, too, grew until the middle of the fourteenth century, achieving sizes that would not be seen again until the nineteenth.

However, so tenuous was this gradual emergence from the long intellectual night, so dependent was it on everything going well, that there were soon setbacks. The climate began to change for the worse in the late thirteenth century, and the cold winters and rainy summers wreaked havoc with crops throughout Europe. In spite of technological improvements, agriculture could not keep up with the growing population. In the early fourteenth century, devastating famines occurred every few years right up to the time of the Black Death.

The result, as always, was the spread of poverty and misery. Most of the towns became overcrowded and filthy. War also played a role in societal breakdown. The Hundred Years War, which broke out in 1337 over the English King Edward III's efforts to claim the kingdom of France, quickly degenerated into an endless series of raids and looting parties. English knights ravaged the French countryside and carried what they could back to England, founding rich dynasties there with their booty while at the same time sowing centuries of hatred between the two countries. All in all, as the historian Barbara Tuchman has vividly pointed out, the fourteenth century was not a good time to be alive.

Things very quickly got much worse. Plague swept through the land of the Uzbeks in central Asia in late 1346. Although the apocalyptic hyperbole that was so beloved of medieval chroniclers makes it difficult to be certain, this may have formed part of an outbreak of the disease that devastated India and China a year or two earlier. As with other natural disasters of the time, tales about the plague were quickly embroidered. Earthquakes, rains of fire and other phenomena were said to have accompanied the onset of disease, but of course the plague itself was bad enough. It soon spread west towards the Black Sea.

In 1346, a small Italian trading colony in the Crimean port of Kaffa was besieged by a Muslim army. When the disease appeared among the besiegers, they catapulted the bodies of its victims over the walls into the city. It is unclear whether this early attempt at germ warfare worked, but either because of this or for other reasons plague soon appeared among the defenders. Remarkably, they were still able to escape from the siege. Some of them, fleeing in ships with their remaining treasures, spread the plague to Constantinople and then to Messina in Sicily. By late 1347, the plague was breaking out everywhere along the Mediterranean coast. We associate the year 1348 with the Black Death because that was the year it spread through Italy, France and England, but in fact the mortality in the Middle East in the preceding two years was at least as devastating.

The plague engendered extreme behaviour. Whole families of victims were nailed up inside their houses and left to starve. Bands of flagellants, endlessly inflicting tortures on themselves, roamed the countryside. These bands started as religious mendicants, but quickly degenerated into a rabble of looters. In many parts of Europe, Jews were accused of bringing on plague by poisoning the water supply, and these baseless rumours started vicious pogroms more terrible than any that were to be seen in Europe before the rise of Hitler. Tens of thousands of Jews were burned to death in cities and towns throughout Spain, Germany, Switzerland and France. These pogroms were sometimes initiated by the flagellants. Although Pope Clement VI and the

FIG. 4-1 Protective clothing worn by a doctor during the Black Death

emperor Charles IV, along with the medical faculties of the universities of Paris and Montpellier, all announced that the Jews were innocent of wrongdoing, their weak protests had no effect because they were not backed by any concrete action. Only King Casimir of Poland managed to prevent pogroms in his country, perhaps because of the entreaties of his Jewish mistress.

Fifty-six thousand people were said to have died of the plague in

Marseilles, and many other towns suffered similarly. Almost everywhere, the first wave of plague killed between twenty per cent and fifty per cent of the people in the communities that it swept through. The mortality was worst in the coastal towns, and indeed some of those that were inland managed to evade its greatest impact. A few city governments instituted sensible regulations. Milan, in particular, vested its board of public health with broad powers and enforced strict sanitation laws. But Milan was an exception.

The plague changed the face of Europe in many ways, both physically and intellectually. In 1348 the city of Siena was about to triple the size of its cathedral, intending that its grandeur would exceed that of St Peter's in Rome. This immense project was stillborn when Siena's population was decimated by the plague. Petrarch's Laura perished, as did many members of the intellectual and artistic élite. And while there was no great upsurge of disbelief in the efficacy of organized religion, like that which would be sparked by the earthquake in Lisbon in 1755, the obvious helplessness of the Church in the face of this apparent act of God began to raise questions that would help to sow the seeds of the Reformation a century and a half later.

The plague spread by both sea and land with amazing rapidity, just as it had 800 years earlier in the time of Justinian. One odd feature, which has puzzled historians of disease ever since, is that it struck down people in remote farms and tiny villages as readily as it did those in the big cities. Even monasteries, those most isolated of communities, were not immune. All the members of an Austin Friars monastery near Avignon and of a Franciscan monastery near Marseilles succumbed.

The Animals' Tale

In 1665 no one was left alive –
In 1666 London was burned to sticks.
Children's rhyme

How could even these isolated communities have been affected so severely? We now know, though medieval scholars and doctors did not, that rats and their fleas, and even on occasion the fleas of humans, are important vectors of the plague, but that it can sometimes spread more directly. Certainly rats must have had a role in the spread of the disease by ship. But it is unclear how much of a role they played in its spread across the land.

Of the two major species of rat associated with humans, the black rat, *Rattus rattus*, is by far the more dangerous. Because it originated in the tropics, when it migrates to the temperate zones it naturally gravitates to warm houses and barns. In addition to plague, it carries typhus, rabies and trichinosis. The brown rat, *Rattus norvegicus*, which despite its name originated in the temperate zone of Asia, rarely carries plague and does not live in close association with humans.

We tend to think of rats as coming to live with our remote ancestors when they were still inhabiting caves. Indeed, a few remains of black rats, dated to 17,000 years ago, have been found in a cave in Bavaria, and the excavators of Egyptian tombs have occasionally discovered the dried bodies of black rats that were accidentally sealed inside. Because black rats are tropical animals, they can only survive in numbers in Europe in the houses, granaries and barns built by humans. These provide them with a precarious niche that protects them from the full force of winter. As such structures were crude and sparse during the Dark Ages, black rats were not particularly common.

At the present time, humans provide rats with an abundance of food and shelter. Brown rats, better adapted to a temperate climate, have largely displaced black rats in the cities of the north. Twentieth-century civilization has proved ideal for rats. There is a growing tendency for the inhabitants of New York and other large cities to eat food in the subways and other public areas. This, coupled with the widespread use of plastic garbage bags instead of metal bins with lids, has given rats greater opportunities than ever. While tales of rats reaching the size of cats are probably exaggerations, the largest brown rat on record weighed 1.6 kilograms and was over half a metre long. Tellingly, urban brown rats tend to be much larger than those from rural areas, suggesting that they have actually been selected for increased size in places where there is plenty of food. The largest rats will be the most successful at competing for mates, and in a food-rich environment they will suffer no disadvantage. Undoubtedly, if there was a rat population explosion in medieval times following the Dark Ages, it must also have been driven by the growing amounts of food as agriculture improved.

Dead and dying black rats are often seen at the beginning of plague outbreaks, as they were in 1994 in the Beed district in India. In both India and China, folk wisdom made a connection between rat plague and human plague long before the discovery of the actual bacillus – when the rats started dying, it was time to flee.

But could black rats, even if they exploded locally into large populations, have accounted for the firestorm of infections during Justinian's Plague and the Black Death? To some authorities this has seemed very doubtful, to the point indeed that it has actually been suggested that neither of these pandemics was bubonic plague at all, but something else like outbreaks of anthrax.

We do not need to go so far – the coincidence of symptoms is too great for either plague to have been anything but bubonic plague. None the less, the question of why the disease spread so rapidly, even in the cold and thinly settled regions of northern Europe with their sparse populations of black rats far from their native tropics, continues to nag at historians and epidemiologists.

Rats and fleas are chancy vectors. When the first cases of plague appear, the human victims have almost certainly acquired the bacteria from rat fleas. The disease, which takes a few days to develop, constitutes the classic bubonic plague with its swellings due to infected lymph nodes. But the plague can spread like wildfire if the bacteria reach the lungs, for there they multiply furiously. Victims cough and spit blood, spewing bacteria into the air with every breath. Anyone near them will breathe in enormous concentrations of bacilli, setting up their own lung infections in turn. So rapid is the progression of this pneumonic form of the disease – untreated victims die in three days, while victims of the bubonic form take five days to die – that the buboes or lymph node swellings that are normally characteristic of the plague have no chance to form. But the plague bacillus does not survive for long outside the body, which means that for even pneumonic plague to spread people must be crowded together. Then why did plague spread so effectively in the countryside as well?

Boccaccio tells an odd tale of his experience of the plague in Florence. He watched two pigs rooting in the discarded rags of a plague victim, and saw them – within minutes – writhe, collapse and die. Later writers have assumed that this event had nothing to do with the plague, but his tale may have been too hastily dismissed. We have had no recent experience with a real plague outbreak – the last substantial ones took

FIG. 4-2 Animal victims of the plague

place in China and India during the first quarter of this century, and while they were disastrous they were not quite as catastrophic as the medieval plagues. We have no idea of what the dynamics of a truly fulminating epidemic might be. During such a plague the bacterial count might reach astronomical proportions in places where people and animals are closely confined. In view of the fact that plague and plague-like bacilli can infect a variety of animals, not just rodents, it may be that a temporary infection of large domestic animals could add fuel to the plague wildfire.

We do not know, after a lapse of centuries, what combination of susceptibilities of humans, rodents, farm animals, and their various fleas might have resulted in the plague that spread with terrible swiftness, not just through the cities and towns, but through the countryside. We do know that a surprising range of animals are susceptible to plague, including dogs and cats (though dogs do not die). When cats are fed infected mice, many of them develop fully-fledged symptoms of plague, complete with buboes.

Animals other than rats may be involved in more recent outbreaks as well. In August 1994, the small 'ratfall' seen by the inhabitants of Mamla, the small village near Beed in Maharashtra state, took place about three weeks before the first human plague victims appeared. Ken Gage, a member of the World Health Organization team that visited the area soon after the outbreak, told me that some cats in the village had died as well – a good indicator that the outbreak was probably plague. The cats had been buried when the WHO team arrived, but they were able to examine several dogs from the villages and found that they had been exposed to the plague bacillus at some point in the past. Unfortunately, of course, we do not know how many of the millions of dogs wandering India's streets and alleys have been similarly exposed.

Our difficulty in imagining the conditions of a fully-fledged plague is complicated by the fact that the plague itself has changed over time. When the descriptions of Procopius, Boccaccio and other eyewitnesses of the earlier plagues are contrasted with the almost casual description of the London plague of 1665 by Samuel Pepys, we are struck by the fact that Pepys seems almost to have been talking about a different disease. The Great Plague of London followed a series of outbreaks of the disease in that city that had occurred intermittently over the previous century, most recently in 1647. At the Great Plague's peak, at the end of August 1665, the Bills of Mortality show that as many as 10,000 of the inhabitants of London had died in a single week. The royal court, and anyone else who could, fled the city. Pepys himself eventually left reluctantly and lived for some weeks near Greenwich. Yet he returned to London as soon as he was able. Even though his own doctor had died of the plague along with many of Pepys's other acquaintances, his own family had been untouched and he was able to carry on his life as if little had happened – celebrating a great naval victory against the Dutch, conducting business in the Royal Exchange and elsewhere, being naughty with an assortment of ladies, and generally behaving with the smug delight in his own accomplishments and growing wealth that we expect of Pepys:

> Thus I ended this month [July 1665] with the greatest joy that ever I did any in my life, because I have spent the greatest part of it with abundance of joy and honour, and pleasant Journeys and brave entertainments, and without cost of money.

After a long spell of hot dry weather broke with heavy rain on 9 September, Pepys noted an immediate decline in the number of plague cases, though he did not connect the probable cleansing effects of the rains with this decline.

In part because of his detachment, Pepys's is one of the very best accounts we have of the plague in London – Daniel Defoe, who was a child at the time, wrote his fictional account *A Journal of the Plague Year* almost sixty years later, and included many incidents that were probably highly embroidered. Pepys does give a few glimpses of the kinds of behaviour people were driven to *in extremis* – recounting for example how people ill with the plague would lean out of their windows and vengefully breathe in the faces of healthy passers-by.

The London plague, dreadful as it was, did not have the full dimensions of horror of the medieval plague three centuries earlier. For Pepys, society continued almost undisturbed. For him and his contemporaries, the toll of the plague was rather like the toll of automobile accidents in society today – regrettable but largely unavoidable.

During the sixteenth and seventeenth centuries, outbreaks of plague in Europe were largely confined to the towns. Repeated waves of plague started every decade or two in London and spread out to the smaller towns over the succeeding two or three years. Sometimes isolated villages, like Eyam in Derbyshire, were devastated as well, but in general simply fleeing into the countryside to avoid the disease, something that had not worked in earlier centuries, began to make a great deal of sense. Isaac Newton did exactly this to avoid the Great Plague of London. And, while the evidence is fragmentary, it seems likely that the mortality due to the plague itself was diminishing, with the outbreaks less severe in the seventeenth than the sixteenth century.

There is another intriguing but fragmentary bit of evidence that the conditions of the disease were changing. The historian Paul Slack and others have observed that the correlation between plague outbreaks and famine, so striking in earlier centuries, had begun to fade by the sixteenth and seventeenth centuries. By that time, the earlier abject and crushing poverty of much of the countryside had been replaced, if not by prosperity, at least by a less desperate way of life. But this does not explain how the earlier waves of the disease spread even to relatively prosperous isolated communities such as monasteries. One is driven to suppose that the way the disease spread might have changed – in a few short centuries, the agent of plague may have evolved. Or, alternatively, the way people lived had changed so much since the fourteenth century that the plague could not spread as easily. Or, very likely, both. My own guess – and it is only a guess – is that the intimate forced association of poverty-stricken farmers with their animals, so widespread during the Dark Ages and medieval times, was beginning to become less common. Perhaps this reduced the number of available paths of infection between animals and humans. If so, then the options open to the plague bacilli had begun to narrow long before the seventeenth century.

Shortly before the Great Plague of 1665, London's Bills of Mortality recorded three outbreaks of plague, in 1630, 1636 and 1647. The most severe, in 1636, killed 10,400. And plague was always present even between outbreaks – only three out of the 22 years for which the Bills survive had no reported cases of plague. Deaths from the plague peaked dramatically in 1665, with 69,596 plague deaths recorded. The numbers dropped precipitously in 1666, to only 1998. This was, of course, the year of the Great Fire, which razed the teeming warrens of the ancient city, burning 'from Pudding Lane to Pie Corner'. Then, astonishingly, the plague quickly vanished. Cases dropped to 35 in 1667, to 14 in 1668, then trickled along sporadically with a few cases each year until 1679, when the last two were recorded. Few other cases have appeared in the city in the succeeding centuries, and all seem to have been introduced by foreign visitors.

Is it too much to suppose that before the Great Fire there was some kind of focus of the plague, perhaps lurking in the sewage-filled alleys and tottering houses of Pudding Lane or nearby Thames Street with its rotting wharves? If there was such a focus, then it was somehow activated every decade or two, spreading its effects out to the rest of the towns of England. The nature of the focus can only be speculated about. Perhaps it was a particularly crowded population of rats and their fleas. If so, then they must have been swarming in a small piece of the medieval world that had lasted beyond its time, somehow surviving down to the London of the Restoration. And it was obviously a piece of that world that the people of London could do without. The Great Fire, widely regarded as a disaster, seems instead to have been a much-needed prophylaxis, burning out a canker in the heart of the city. And once that canker disappeared, the cycle of the plague bacillus was broken as well.

Yet if the Great Fire had burned out the only source of the plague, one would expect that the disease would have disappeared immediately. There must have been other sources that gave rise to the few cases that followed the Great Fire, and that were removed as the city gradually left behind the filth and squalor of medieval times.

The plague faded away in the rest of Europe at about the same time, though not all at once and not correlated with so obvious an event as London's Great Fire. The last major outbreak in Italy, involving regions to the north and south of Rome as well as the city itself, took place in 1656. In France, there was a severe outbreak in the town of Amiens in 1667–8, but oddly none took place in the much more crowded city of Paris. The final French flare-up occurred in Marseilles in 1720, killing 50,000 of its inhabitants, though this was probably an introduction from abroad. Plague had repeatedly devastated the entire Iberian peninsula during the sixteenth and seventeenth centuries, with the most massive outbreak taking place between 1647 and 1652. It contributed greatly to the political upheavals of the time. A less severe outbreak, and the last in the region, took place between 1676 and 1685. And finally there was a severe epidemic in Hungary between 1739 and 1742, the last major one in Europe.

Something was happening during this time throughout Europe, even in its most backward regions, something that broke the cycle of the plague with surprising ease. Why was the plague's grip so tenuous that it could be defeated by almost imperceptible changes in the ways that people lived?

The Bacillus' Tale

Discovery of the agent of plague had to await the remarkable advances of nineteenth-century biological science. Bacteria are tiny rods, spheres and spirals, usually almost colourless, far smaller than the cells of their hosts. Their virtual invisibility even under magnification explains why they remained unnoticed for so long in spite of intense scrutiny of both healthy and diseased tissues by many microscopists of the seventeenth and eighteenth centuries. Their discovery had to await improvements in microscope optics and the invention of specific stains that could be used to colour the bacteria. Once they were found, the next important

FIG. 4-3 The Black Death in Naples, 1656

advance was the realization that these micro-organisms could be grown separately from their hosts in pure culture and reinoculated into new animals. If the same disease resulted from such reinoculation and the organism could be isolated from these animals in turn, then, as the microbiologist Robert Koch pointed out, the likelihood was strong that the causative agent had been found.

The plague bacillus was discovered in the course of the great epidemic that swept through southern China in 1894. Two scientists claimed to have found it, and for a long time there was great confusion about which of them was right. Much of the confusion stemmed from the fact that the microbiological community was split at the end of the nineteenth century into two intensely nationalistic factions, the followers of the Frenchman Louis Pasteur and those of the German Robert Koch.

Alexandre Yersin, a young Swiss medical student who had trained at the Institut Pasteur, was a partisan of the French school. When he began his hunt for the plague bacterium during the last decade of the nineteenth century, he already had impeccable credentials for the search, though he did not realize how impeccable they were. In the late 1880s, working at Pasteur's Institute in Paris, he had succeeded in isolating a bacterium from guinea pigs and rabbits which were suffering from a kind of animal tuberculosis. He showed, using the reasoning that had been pioneered by Koch, that inoculation of healthy animals with this bacillus would give them the disease. The bacillus is now known as *Yersinia pseudotuberculosis*, in Yersin's honour, and it is the same one that killed those muskoxen in the remote northern reaches of Canada. Yersin could not have known that it is an extremely close relative – unnervingly close – of the bacillus that causes bubonic plague.

Although a brilliant career in medicine beckoned, Yersin was then seized by wanderlust. Abandoning his family and friends, he set off for Southeast Asia and a completely different kind of career. He arrived in the part of Indochina that is now Vietnam, and immediately embarked on a series of mapping explorations of the unknown mountain ranges of the interior. It turned out that he was as good a mapmaker as he was a bacteriologist. With Pasteur's help, Yersin obtained some support for his cartographic work from French geographical societies. But he could not avoid getting involved again with disease, for everywhere he went he found cases of the plague. The closer he got to the Chinese border, the commoner the disease became. He wanted to trace it to its source, which he suspected to be adjacent to the northern parts of the province of Tonkin, in the Chinese province of Yunnan, but the colonial governor refused him permission – presumably for fear of what he would find. The governor's reply to his entreaties was: 'There has never been any plague in Yunnan, and if there were I would deny it.'

It soon became obvious that there was indeed plague in Yunnan, however. In 1894 a huge outbreak occurred in the city of Canton, killing perhaps 100,000 people. And the plague soon spread to nearby Hong Kong, where it could no longer be ignored. Yersin pleaded to be sent there, and eventually gained permission, but he was given no official backing.

Plentiful official backing had, however, been provided to Shibasaburo Kitasato, the representative of the Koch school who arrived in Hong Kong at the same time. While he was with Koch in Germany, Kitasato had done important work on vaccines for tetanus and diphtheria. He was also the co-discoverer of antibodies, the proteins that make up a central part of the immune system. He arrived in the city with a large retinue of assistants and mountains of equipment. Welcomed at Hong Kong's hospitals, he swiftly cultured a bacterium from the finger of an autopsied plague victim, then announced that he had found the bacillus of plague.

Yersin was denied access to the hospitals, and indeed during the few times that he and Kitasato met their relationship could be charitably described as one of strained politeness. To obtain material he was forced to bribe the English sailors who were in charge of burying some of the victims. They let him into a cellar where the bodies were kept for a few hours before burial, resting in rough coffins on a bed of lime.

Yersin took a more careful approach than Kitasato's. He cultured his candidate bacillus from the inflamed lymph nodes of the victims, the buboes themselves, which he found to be filled with bacteria that looked like tiny fat ovals. He quickly found that the bacilli were gram-negative – that is, when they were stained with a concoction of crystal violet called Gram stain and then killed, most of the stain could easily be washed out from the dead cells.

Kitasato's bacillus, rounder and fatter, was a coccus, and when he stained it with crystal violet it retained the stain. It was gram-positive.

Robert Koch had clearly set down four rules for determining the causative agent of a disease – his famous postulates. It must be found in the victim, preferably at the site of the disease itself – Yersin's bacillus fit this criterion more closely than Kitasato's, although both strains had certainly come from plague victims. It must be culturable to form a pure strain. Then, when it is inoculated into a healthy host, the new host must come down with the disease. Finally, the presumptive causative agent must then be culturable from the new host. These last two steps were of course impossible to carry out in humans, but Yersin was able to inject his bacillus into rats. It quickly killed them, and he then isolated the same bacillus in large numbers from the dead rats. Kitasato was never able to close this circle of proof using his coccus.

Returning to France, Yersin inoculated his bacillus into horses to produce an antiserum. During another outbreak of the plague in Hong Kong in 1896, he returned and cured several victims with it – the first successful cures for the plague in the history of the world.

After decades of controversy, it is now universally agreed that Yersin and not Kitasato was the discoverer of the plague bacillus. Yersin originally called it *Pasteurella pestis* in honor of his mentor, but it has since been found to be much more closely related to his earlier bacillus that caused tuberculosis in animals than it is to other bacteria of the *Pasteurella* group. It has been renamed *Yersinia pestis*.

Yersin spent most of the rest of his life in Indochina, training doctors from among the local people and defending them against the stupidities of the colonial administration. He also helped to found the rubber industry in the area, and imported cinchona trees for the production of the quinine that was the only specific treatment at the time for malaria. By his death in 1943 he had become a much-admired figure. But during his work in Hong Kong he had missed an important step in understanding how the plague was spread, for he knew nothing about the role of rats or their fleas. This gap was filled in by the physician Masanori Ogata, working in Formosa, and independently by the medical missionary Paul Louis Simond in Bombay. Both observed that rats often came down with symptoms very much like those of human plague, complete with swollen lymph nodes that looked like the characteristic buboes of humans. Further, they found that human victims of the plague had often been bitten by fleas. Both tried some simple experiments to see whether rat fleas might be involved in the disease's spread. They ground up fleas and injected them into healthy rats, and found that the rats quickly developed the unmistakable symptoms of plague. Simond went a step further and discovered that massive plugs of plague bacilli were blocking the guts of the infected fleas. The work of these pioneers was dismissed for some years, until it was finally confirmed in 1905 during a subsequent outbreak of plague in India.

With this discovery, it seemed that the emerging science of microbiology had added another shining prize to its string of triumphs. The causative agent for plague had certainly been found. A successful serum against it had been produced, and the missing link in the spread of the disease, the rat flea, had also been discovered. But the complexity of the story had hardly begun to be explored. The bacteria themselves were still an enigma.

Much of what distinguishes dangerous from benign bacteria revolves around how they invade cells. There is an eerie elegance in the way that plague bacteria and the various kinds of cells that make up the

FIG. 4-4 A. Louis Pasteur;
B. Robert Koch and
Shibasaburo Kitasato;
C. Alexandre Yersin

tissues of their hosts interact with each other, each always exploiting the other's weakness.

Yersinia pestis is one of ten species that have been distinguished in the genus *Yersinia*, all closely related. The actual number of species is a bit of a guess – it is much harder to tell various species of bacteria apart than it is to distinguish species of groups such as birds or insects. This is because species of bacteria have far fewer obvious properties that can define them, and often these properties are simple ones like the ability to grow on a certain sugar, to make a particular amino acid, or to elicit a specific antibody in their hosts. Two species of bacteria need not be as distinct as two species of hummingbird, and sometimes bacteriologists discover to their surprise that only a few mutational changes or gene-borrowings are enough to turn one species into something almost identical to another.

Three of the *Yersinia* are pathogenic in animals: *Y. pseudotuberculosis*, the bacterium that Yersin had found back in France, another species called *Y. enterocolitica*, and of course the plague bacillus *Y. pestis*. The first two are ubiquitous in the environment, far commoner than *Y. pestis*. They are often found in wild and domesticated animals. They are also, unlike *Y. pestis*, motile – they can swim about in an aqueous medium using long thin flagella. Just like their more remote relative *Salmonella*, they can enter our digestive tracts through contaminated food. Once there they quickly invade the cells lining the inner walls of the intestines. Sometimes they have little effect, but sometimes, just as with *Salmonella*, severe cases of *Yersinia* infection can cause vomiting and diarrhoea. Occasionally, rampant infection can result in tissue destruction and even death. And if the lungs are infected, the destruction can be very like that caused by tuberculosis. Occasionally, as we saw with the muskoxen, there are outbreaks of *Yersinia* infections that are just as deadly to their animal hosts as plague is to humans.

Ralph Isberg and Stanley Falkow of Stanford University were the first to discover exactly how *Y. pseudotuberculosis* enters the cells of its host. The bacterium manufactures a protein called invasin, molecules of which migrate to its surface in large numbers. Invasin forms a strong attachment to other specific molecules, known as integrins, that have accumulated by a similar process on the surfaces of the host's gut cells. (Integrins are involved in various functions in the host cell, and the bacterial invasin has evolved the ability to cling to these highly visible proteins.) Once a bacterium covered with invasin molecules attaches to a cell of its host, the cell membrane breaks down at the point of attachment and the bacterium can slip inside. But it must do so quickly, before the patrolling white blood cells called macrophages or phagocytes of the host's immune system arrive at the site of the bacterial invasion.

These prowling phagocytes are also covered with a variety of protein molecules that are specialized for different functions. Some of these proteins are the same integrins that the bacteria can attach to. The result is that if the phagocytes arrive quickly enough on the scene, they too readily bind to the bacteria and ingest them – but the crucial difference is that the phagocytes can destroy the bacteria once they have engulfed them, while the gut cells cannot.

In order to survive in its host, *Y. pseudotuberculosis* must therefore be very good at entering gut cells, so good that an appreciable number can hide there before the phagocytes can find them. Usually this hide-and-seek process has few consequences for the host. Once they are safely hidden, the bacteria have a good chance of spreading to other members of the host species, because the cells that they infect are continually being shed from the gut lining. The cells pass out in the faeces, and the bacteria they contain become part of the teeming bacterial population of the soil. There they are able to persist long enough to be picked up by another host animal.

They can increase their chances of being passed on, of course, if they multiply to high numbers in the host, but here is where the game can turn deadly. Strains of *Yersinia* that multiply in the gut in large numbers can cause lesions in the gut lining, which spew out more bacteria and their accompanying toxins. The host animal may vomit and suffer severe diarrhoea. All this upheaval means that far more bacteria can escape into the environment, but at a cost. There is a high probability that this violent reaction will kill the host, since a sick animal is unlikely to survive for long in the wild. In effect, the bacteria have given their host a mild kind of plague, briefly increasing their own chances of survival but only at the host's expense.

Strains of bacteria that take this more risky route are arising all the time. This can happen by mutation, or they may pick up an odd bit of DNA from a kind of dilute brew of DNA molecules that is always present in the host's gut or in the soil. These DNA molecules often come from other, completely different, species of bacteria. They may

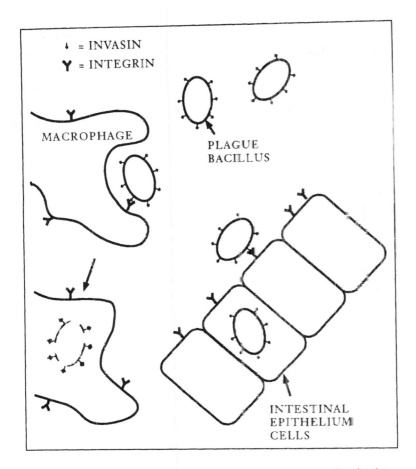

FIG. 4-5 The invasin molecules on the surface of the *Yersinia* bacilli that invade a human host can bind to integrins on the surface of intestinal epithelium cells. The bacteria can then pass safely through the cell membranes into the interiors of the cells, where they are able to survive. But similar integrins on the surface of the patrolling white blood cells called macrophages can become attached to the bacterial invasins. When the bacteria are engulfed by the macrophages, they are rapidly digested.

have escaped from various bacterial cells that have died and broken open, or they may be packaged inside viruses called bacteriophages that are specialized to live on bacteria. *Yersinia* and other bacteria can then pick them up, as if they were sale items in a bargain basement.

If these little bits of DNA confer sufficient virulence, bacteria that acquire them may be aided in their ability to infect the host animal and multiply. But these mutant bacteria are unlikely to survive unless there are many host animals nearby, enabling them to spread even if their immediate host dies as a result of their activities. If this condition is met the result can be an animal plague or *epizootic* (as distinct from a human epidemic). Most such epizootic plagues pass unnoticed because the animals that are affected tend to be small and their demise is hidden from our view. But sometimes, if the animals happen to be large like those muskoxen that died on the Canadian tundra, the effects can be dramatic.

No Boccaccio, Defoe or Pepys has recorded the billions of sad little deaths that have taken place as a result of these animal plagues. The end product of this long grim history, however, has been the evolution of strains of *Yersinia* that are capable of producing a great range of types of infection, all the way from mild and almost unnoticeable invasions of a few gut cells to the explosive and fully-fledged manifestations of the Black Death.

For many years this great variety of responses and abilities tended to cause confusion among microbiologists who worked with *Yersinia*. Strains isolated from nature were often virulent in laboratory animals, but after they were cultured for a few generations their virulence would be lost. It was not until new ways of examining their DNA became available that it was discovered that their virulence resided chiefly in two small 'extra' chromosomes called plasmids.

A bacterium normally has one large chromosome, in the form of a long double-helical strand of DNA that is joined at its ends to form an immense circle. Often, however, it may carry smaller circles of DNA that are not necessary for its survival but which may contain genes not found on the main chromosome. If a cell carrying such a plasmid happens to break open and release its DNA, then the plasmid can actually be picked up by another species of bacterium – the bargain basement effect. The most virulent strains of *Yersinia*, including *Y. pestis*, carry these plasmids, which provide them with extra capabilities. But often, particularly when the bacteria are cultured in the laboratory, the plasmids are lost, and so is the virulence.

The plague that affects humans is so severe, so overwhelming in its symptoms, that one is tempted to suppose that *Y. pestis* must somehow have acquired even more virulence factors than its less fatal brethren. Actually, the opposite is the case. It turns out that the cause of human plague is a crippled bacterium that has had many of its options for escape to a new host closed to it. But, like a wounded animal backed into a corner, *Y. pestis* can still strike back using all its remaining powers.

The tally of these missing capabilities of *Yersinia pestis* became longer and longer the more it was studied. To begin with, as we saw earlier, it has lost the ability to swim about, and must drift passively. In addition, it has lost so many biochemical pathways as a result of mutation that it is unable to survive for long outside the body of its animal host. This means that if *Y. pestis* is carried in faeces from the animal to the soil, it will soon die. One of its most startling defects is a missing step in an absolutely essential part of cellular metabolism known as the Krebs Cycle. This cycle of biochemical reactions is a source of building blocks that are needed to make amino acids and compounds essential for respiration. It is so important that its discovery won the chemist Hans Krebs the Nobel Prize. The *Y. pestis* bacteria that lack part of this cycle are unable to survive for long unless they can obtain the missing compounds from their host. The host might be a human, a rat or a flea, but a host is essential. This is an important reason why these bacteria cannot survive in the soil.

The existence of many of these defects had been known for years. Then, in 1987, Daniel Sikkema and Robert Brubaker of Michigan State University reported the discovery of an even more startling defect, one that has large implications for virulence. They found that, unlike its free-living relatives, *Y. pestis* actually lacks the ability to invade cultures of human cells grown in the laboratory. The defect is connected to a difficulty in taking up iron-containing compounds that it needs to survive. The weakened bacillus must survive by drifting about and obtaining its nourishment from the circulatory fluids.

This inability to hide in the cells of their host should leave the bacteria helpless against attacks from the host's patrolling phagocytes. And if phagocytes are able to catch and eat these poor helpless mutant bacteria, then surely this ought to make them less virulent, not more. Instead, these crippled bacteria actually turn out to be far more virulent than their relatives that are still able to hide in the cells of their hosts.

This discovery spawned a flurry of research. What were the really important genetic differences between *Y. pestis* and its less virulent relatives? Some of the differences were tracked down by scientists working at the University of Umeå in Sweden, 500 kilometres north of

Stockholm on the shores of the frigid Gulf of Bothnia. In that remote place, Hans Wolf-Watz and his colleagues have been experimenting for years with both *Y. pseudotuberculosis* and *Y. pestis*, trying to pin down just what distinguishes them.

At the outset, they found that *Y. pestis* did not even make the invasin protein of Isberg and Falkow, which seemed to be one obvious reason why it could not invade host cells. So they next asked what would happen if they destroyed the invasin gene in *Y. pseudotuberculosis*. Their idea was to try to turn it into a kind of imitation *Y. pestis*. It was straightforward for them, using modern molecular biological methods, to destroy this specific gene on the bacterial chromosome, leaving all the rest of the genes intact. They found that *Y. pseudotuberculosis* that had been crippled in this way could no longer invade cultured human cells, so that now it acted more like *Y. pestis*. Other properties had changed too, for the altered bacteria had suddenly become more virulent – it took far fewer of the crippled *Y. pseudotuberculosis* to kill mice.

While their modified *Y. pseudotuberculosis* was nowhere near as virulent as *Y. pestis*, they had retraced in the laboratory at least part of the evolutionary path that separated the two bacterial species. And this had been done by destroying a gene rather than by introducing some new virulence factor.

They then went a step further and destroyed another gene that they suspected from previous work might have an invasin-like function. This gene was found not on the main chromosome but on one of the plasmids. When this second gene was destroyed, the virulence of the crippled bacteria increased ten thousandfold when they were injected beneath the skin of mice, and a thousandfold when they were given orally. These doubly crippled bacteria were now behaving much more like their fierce relatives, *Y. pestis*.

Wolf-Watz was able to find and clone the equivalent gene in *Y. pestis*, which was also harboured in a plasmid. He found that this gene, too, had been damaged so that it could no longer work. *Y. pestis*, then, just like their laboratory strain of *Y. pseudotuberculosis*, was doubly crippled. The remarkable thing, of course, was that the defective *Y. pestis* strains had been isolated not from the laboratory but from the real world. The crippling defects of *Y. pestis* had arisen in nature, at some unknown point in the past.

Wolf-Watz and his colleagues had been able to imitate this double genetic blow by deliberate genetic manipulation of its close relative *Y. pseudotuberculosis*. They had been able to retrace part of the evolutionary pathway that separated *Y. pseudotuberculosis* from *Y. pestis*, a pathway that had already been followed long before in the natural world.

Thus, it appears that some of the genetic differences that separate *Y. pestis* from *Y. pseudotuberculosis* are rather simple. It is easy to make a mutation in a gene that will destroy its function, and such mutations are likely to arise very commonly in the natural world as well as in the laboratory. Because of this, Wolf-Watz suggested that there is a real possibility that *Y. pestis* might have arisen more than once, perhaps from *Y. pseudotuberculosis* or from some other close relative. Perhaps strains of *Y. pestis* might have arisen independently centuries ago in both Asia and Africa, giving rise to those two great plagues that bracketed the Dark Ages.

On the face of it, such an evolutionary situation ought to be highly unlikely. Defective genes arise in natural populations all the time, but only rarely and under unusual circumstances do they spread through an entire population. In order for this to happen, the descendants of the organism in which the defect arose must somehow out-reproduce all the other members of the population, so that the harmful gene will be able to spread. This is particularly improbable if the population is large, as it is in bacteria. And you can see immediately that the more harmful the mutation, the less likely it is to spread in such a fashion.

While it is possible to imagine, barely, that one such mutation might have arisen and spread through the *Y. pestis* population by chance, the likelihood that two independent mutations could have done so is far smaller. The difficulty is compounded when we remember all the other mutations that have helped to make *Y. pestis* so defective, like the one that damaged the Krebs Cycle. There has to be some good reason why these mutant strains of bacteria have been so successful, even though they appear to have decreased capabilities.

More recently, other parts of the story have emerged. Thomas Quan and a group of colleagues at the Centers for Disease Control in Fort Collins, Colorado, have managed to destroy another gene on one of the plasmids that is carried by *Y. pestis*. This gene codes for a protease, which is an enzyme that digests other proteins. They found that *Y. pestis* carrying the destroyed gene were a million times *less* virulent than those that carried the intact gene. So this protease gene contributes to the virulence of *Y. pestis* when it is functional, not when it is mutant, which is quite the opposite of the effects of the genes investigated by Wolf-Watz.

Quan and his colleagues therefore thought that Wolf-Watz must be wrong, that it is not the loss of genes that contributes to the viciousness of *Y. pestis* but rather the acquisition from some unrelated bacterium of a plasmid that carries the protease. Searching among bacteria that are close relatives of *Yersinia*, they found that similar plasmids could be found in *Salmonella* and even in the common gut bacterium *E. coli*. Perhaps, they thought, the virulent *Yersinia* had acquired its protease-gene-carrying plasmid from one of these bacteria.

The two groups of scientists, I suspect, are both right. They have uncovered different parts of the story. There is no doubt that, as Wolf-Watz found, *Y. pestis* is damaged genetically, increasing its virulence. On the other hand, Quan has certainly managed to show that a gene on one of the plasmids it has picked up also helps it to invade and multiply within its host. Remember, though, that the plasmid can be acquired in a single step, so that although *Y. pestis* has gained some of its virulence from a mutation that happened in its invasin gene and more virulence from a plasmid carrying a protease gene, after all this it is still a few simple genetic steps away from being harmless.

If *Y. pestis* is such a terrifying pathogen, then how has it managed to paint itself into such a corner? Why has it acquired all these genetic defects that prevent it from living anywhere except in the host's body? Its vulnerable genotype tells us that it is forced to live in animal hosts most of the time, and yet it rapidly kills those hosts. It can only be transmitted to new hosts if the host population is numerous enough, and if fleas happen to be available. But if hosts are sparse and fleas are few, than it will kill its host too soon and find itself trapped in a rapidly cooling dead body with no means of escape.

The plague bacillus cannot live anywhere but inside its hosts, but it is so fragile that most of the time it must be cautious. If it behaves with wild abandon and causes the plague, then most of the time it will die as well. To explain all the genetic abilities that *Y. pestis* has lost, we are forced to conclude that most of the time, for whatever reason, it must refrain from causing the plague. Most of the time, *Y. pestis* cannot be – indeed, cannot afford to be – a plague-causing bacillus.

There are, therefore, likely to be less virulent close relatives of *Y. pestis* lurking in our penumbra of diseases. We must remember that the highly virulent laboratory strains of *Y. pestis* were isolated from humans and animals that had actually come down with the plague, and as a result they are hardly typical. They are likely to be a small subset of the universe of *Y. pestis*-like bacteria that are scattered throughout the soil, the water, and the gastrointestinal tracts of many different animals including ourselves.

Another possibility is that *Y. pestis* itself, with all its defects. lives in other animals, but does not for some reason cause the dramatic symptoms of plague. Recent experiments by Quan and his co-workers reinforce this possibility – they deliberately infected black-footed ferrets and Siberian polecats, likely carriers of the plague bacillus, with a highly virulent strain of *Y. pestis* and found that it caused no symptoms. Remarkably, they also discovered that wild ferrets trapped from a Wyoming population were already making high levels of antibodies to *Y. pestis*, which means that they must have been exposed to this bacillus (or a close relative) while they were living in the wild. If so, since *Y. pestis* cannot live in the soil, these strains must be passed directly from animal to animal.

The crippled *Y. pestis* might never have evolved if its relative *Y. pseudotuberculosis* were itself only one or two genetic steps away from being able to cause the plague in humans. This robust organism *can* live in the soil, and *can* be passed easily from one host to another even if the population is sparse. It can occasionally cause something very like the horror of bubonic plague in other animals. So why is the wimpish, genetically damaged *Y. pestis* so much more potentially dangerous to us than the strong *Y. pseudotuberculosis* that has all its faculties intact?

To get at least a glimpse of an answer to this question, we must follow the history of *Y. pestis* in the bodies of its victims, and see how it manages to pass from one human to another.

The Flea's Tale

When *Y. pestis* invades the body of a human or animal host, its defective genes prevent it from entering the host's cells. The bacteria are therefore exposed to the phagocytes that should be able to hunt and destroy them. But they are not helpless victims of these cruising defenders

They have a weapon against them, in the form of the protease that Quan and his colleagues discovered.

Whenever the bacteria come in contact with phagocytes, the protease destroys or damages proteins on the phagocytes' surface. So long as their protease gene is intact, the bacteria can ward off the phagocytes even though they cannot get inside the host's cells. This allows them to multiply relatively unhampered inside the host's body. During a full-fledged Y. *pestis* infection the bacteria are free to go gather in great clots under the skin and accumulate in the lymph nodes, because the immune system is helpless to destroy them. The result is the black blotches and swollen buboes of the Black Death.

It is the sheer numbers of these bacteria that kill their hosts, by inducing septic shock. They stimulate the host's immune system in a nonspecific and uncontrolled way, producing high fever and choking swelling of the throat.

So immense do the bacterial numbers become that they often manage to invade the lungs, where they immediately begin to break down the delicate tissues. As the lungs literally dissolve under this onslaught, the bacteria are broadcast on the breath, in tiny droplets of fluid. This bacterial destruction of the lungs accounts for the fetid breath of plague victims that figures in all the medieval accounts of the pestilence. And it also accounts for the sudden emergence of pneumonic plague during a bubonic plague infection. Stanley Falkow of Stanford University, who has looked into many old autopsy reports, tells me that about half the victims of bubonic plague show invasion of the lung tissue. Pneumonic plague, it seems, is only a short but catastrophic step away from the bubonic variety. It probably does not require any further mutation of the plague bacilli.

So devastating is the infection that the human or animal host will almost certainly soon die—and when this happens the invading bacteria will die as well. Their immense numbers, however, are their temporary salvation. The great gatherings of bacteria under the skin are easily picked up by fleas before the host dies and are transmitted to new hosts, spreading the disease and giving the bacteria temporary homes before their rampage destroys them. Or, in the pneumonic form of the plague, the bacteria are simply spread to new hosts through the air.

This story of bacteria run amok turns out to be even more complex, for up to now we have only considered their effect on their human or animal host. They also affect the fleas of both rats and humans in such a way as to aid their own spread. The fleas are actually unusual hosts for *Yersinia* bacteria, only becoming important when the bacteria are prevented from spreading through the faeces and the soil.

When the fleas ingest large numbers of bacteria from their hosts they too become ill, and their illness helps to spread the disease. You will remember that the first scientists to work with rat fleas found that their guts were often clogged with masses of multiplying bacteria. As a result the fleas could not absorb moisture from their blood meals. They became progressively more and more thirsty, jumping from host to host in a desperate attempt to find moisture. They had been converted into living hypodermic needles, injecting the bacteria into the skins of their new hosts whenever they tried to quench their thirst.

Why should the bacteria be able to affect the fleas so dramatically? The fleas are not defenceless. Although they and other insects do not have the elaborate immune systems that are found in mammals and birds, they do have patrolling phagocytes that can destroy bacteria and other invaders. Indeed, the first phagocytes that were ever seen under the microscope belonged to water fleas, remote (but not too remote) relatives of rat and human fleas. Early in this century, the Russian scientist Elie Metchnikoff observed the phagocytes of these tiny crustacea engulfing bits of foreign matter. In theory, then, the rat or human fleas should be able to destroy the clots of bacteria in their guts, restoring them to normal working order.

However – and this is speculation – it may not be a coincidence that *Y. pestis* can multiply so readily in the bodies of fleas. Perhaps they can employ the same proteases that they use to damage human phagocytes against the phagocytes of fleas. This would make the fleas' phagocytes unable to engulf the invaders. It may be – although there is as yet no evidence – that the multiplication of bacteria in fleas as well as in humans and rats is aided by an inability to destroy the invaders.

There are also strong indications that the bacteria have been in their flea hosts many times before, and have reached a complex genetic adjustment with them. Although they must be able to make the fleas desperately hungry and thirsty, they cannot kill them too quickly, or they will not be carried to new mammalian hosts. This fine-tuning of their effects on the fleas is the result of a complex series of events.

Even before the bacteria multiply to large numbers in the fleas, they can immediately block their guts. Blood taken in by the fleas is rapidly clotted by a protein that is manufactured by the bacteria ingested at the same time. The fleas soon begin to starve and dehydrate. But before they can die, another protein begins to break down the newly-formed blood clots. The genes for both proteins are carried by a plasmid that is unique to *Y. pestis*, a plasmid that enables it to make the most out of its short existence in the infected fleas.

When Quan and his colleagues deliberately infected fleas with bacteria in which both of the genes had been destroyed, they died more rapidly than fleas that had been infected by intact *Y. pestis*. It seems that the blood-clotting proteins may not be as important as the protein that breaks down the clot once it forms. A little bit of breakdown of the clot of blood and bacteria in the gut of the flea may be enough to ensure that the flea will gain a little nourishment. At the same time, bits of the clot will break free and travel to the fleas' mouth parts – a kind of insect heartburn. All these elaborate events seem to be necessary because the bacteria cannot swim there on their own – you will recall that they have lost their ability to be motile. So *Y. pestis* has been aided in coping with the difficult situation that it must confront in the bodies of fleas by the acquisition of yet another plasmid, one that just happens to be around and just happens to be excellent at the task.

The fleas, of course, are nothing more than mobile hypodermic needles. Virtually any kind of flea can be invaded by the bacteria, if it happens to have fed on animals that have large numbers of bacteria in their blood. As a result, these sick bacteria can infect mice, rats, cats, perhaps farm animals, and of course humans. (It may have been more than just an unreasoning fear of witchcraft that explains why so many cats were hunted down and killed during the Black Death, since these animals are highly susceptible to the plague.)

We now have many pieces of the puzzle. Rare as bubonic plague outbreaks are, they have still been shaped to a remarkable extent by evolution – even though it is a particularly crude and untidy sort of evolution. *Y. pestis* bacilli have acquired a number of mutations that block critical functions. At the same time, they have picked up plasmids that allow them to acquire other functions. These abilities protect them from the hosts' immune systems or allow them to survive in the fleas. Provided that the plasmids are available in the bacterial environment (and goodness knows what organisms they evolved in originally), it is presumably fairly easy for such bits and pieces to come together to make a nasty opportunist like *Y. pestis* – a kind of evolutionary Frankenstein's monster.

It is one of nature's supreme ironies that, during the unusual juxtaposition of circumstances that leads to an outbreak of plague, all the major participants in the drama are sick. This includes the humans, the rats, the fleas, and the very bacteria themselves. And this may help to explain why, even in the filthy and squalid medieval world, severe outbreaks of plague were relatively rare. No ecological relationship in which all the interacting members are behaving abnormally is likely to persist for long.

Improved sanitation is the key to preventing plague outbreaks, even in situations where a few decades ago plague would have been a certainty. Parts of Hong Kong, where Alexandre Yersin cured the first cases of the plague in 1896, now have the highest population densities in the world. The Shamshuipo district of Kowloon has 16,000 people per square kilometre, and a single housing complex can contain 100,000 people. Yet plague in this clean, well-run city is now almost unknown.

Even though a few cases of plague, mostly from contact with wild rodents, appear every year in the US, we tend to think of the disease as being a thing of the past. These occasional cases, if they are diagnosed early enough, can now usually be controlled by antibiotics (though as we have seen, mortality even with the best medical care is still about fifteen per cent). And while the population of rats has certainly grown since medieval times (as have the rats themselves), we mostly keep them at arm's length – although they can still swarm over the steps of the New York Public Library at dusk, driving away hysterical tourists. But until recently we supposed that the unusual circumstances that engender plague had receded into history.

In fact, history may have caught up with us. Plague bacilli and their close relatives have certainly now escaped from their original foci in Africa and Asia, and have been carried to every continent. The sheer mass of our enteric bacteria, and the changing mix of bacteria in the soil and sewage that surround human population centres, particularly in the Third World, are something that has never been seen on the planet before. The niches available to these bacilli are more numerous than ever. Potential agents of the plague are everywhere in our food and in the soil.

5

Was the Indian plague actually plague, and if not why not?

Litter Bringeth Plague!
Admonitory sign at a California Renaissance Faire

Madras is a jumbled tropical city of five and a half million, flanked by a magnificent but polluted beach that stretches for a hundred unbroken kilometres along the Bay of Bengal. It was my point of arrival in the Indian subcontinent, and I quickly learned the first rule of survival in the city. If I glimpsed a stretch of open water, I held my breath and headed away from it as quickly as possible.

Open water filled with sewage is everywhere in Madras. The River Cooum oozes its unspeakable way through the heart of the city, and the River Adyar, only marginally less horrific, flanks it to the south. Equally malodorous lakes and swamps abound, most of them surrounded by shacks and shanties, and swarming with people bathing and washing their clothes in an effort to remove – or at least dilute – the pervasive filth. Everywhere there are rotting mounds of garbage, so repellent that they are spurned even by the goats.

As I travelled around India, I talked to dozens of scientists and bureaucrats about dozens of diseases. These ranged from major killers such as dysentery, TB and pneumonia to the more exotic chikungunya and Japanese encephalitis. Yet everywhere I went, one disease was inevitably the first topic of discussion. That disease was plague, the apparent re-emergence of which had petrified the world in September 1994. As I tried to follow up the story of the outbreak, I found myself enmeshed in a tangle of politics and science, confronted with as many opinions as people. I finished the trip in a state of sceptical confusion. Had there been an outbreak of plague? And if, as many Indian scientists claimed, it wasn't plague, then why had the disease not broken out when conditions seemed to be so ideal for it?

When I first encountered Madras, it struck me as being an ideal breeding ground for any number of diseases. It did not surprise me to learn that the city was one of the places where the last sporadic cases of bubonic plague before the 1994 outbreak had been reported. Between 1960 and 1968 a small number of cases had appeared in Madras, along with a few in Mysore, Himachal Pradesh and Rajasthan. These numbers were even fewer than those seen in California during the same period. It seemed at the time that they were the last few sputterings of this dreadful disease, the end of its long and grim history on the subcontinent.

Then came September 1994, with the outbreaks in the Beed district and Surat. These were on the other side of the country from Madras, in areas where plague had not been reported since the 1940s. At the end of the first week of the outbreak in Surat, about fifty people had died from acute respiratory distress, but the disease (whatever it was) had quickly been brought under control. The government congratulated itself that it had prevented a full-blown epidemic.

The World Health Organization team that arrived in October had split up into three groups. These groups of scientists spent ten busy days in Surat, Delhi and the Beed area, doing epidemiological studies and taking samples. They obtained what appeared to be some positive traces of antibody to the bacteria in the blood of some of the victims – five out of 44 in Surat, and a few apparent positives in Beed.

The Beed samples were the most clear-cut, and even those were somewhat questionable. A positive diagnosis requires that paired samples be taken, one during the early stages of infection and one during convalescence, to see if the antibody titre has risen during the illness. Four sets of specimens seemed to show such a rise, but unfortunately in the confusion it was unclear whether the samples had been properly labelled.

Most frustratingly, none of the teams' efforts at culturing the bacteria was successful, because of contamination of the culture medium. And the Indian government did not allow samples to be taken out of the country for testing by WHO laboratories. But in spite of all the difficulties and bureaucratic hurdles, the WHO report that resulted from the visit agreed with the Indian government. On the basis of the apparently positive blood samples, it seemed that there had indeed been a plague outbreak, though its true extent was unclear.

Yet the members of the WHO teams were very puzzled by the fact that nobody appeared to have died in the Beed area. It was possible that there had been deaths that had gone unreported, and I was told in early 1995 that investigations were continuing.

Indian scientists were puzzled as well. In November 1994, two letters appeared in the *Lancet*, one from a group at the All India Institute of Medical Sciences in New Delhi and the other from T. Jacob John, the head of the microbiology department at the Christian Medical College in Vellore in southern India. Both questioned whether the outbreaks could have been plague, particularly the outbreak in Surat.

Jacob John was the first scientist I talked to about the plague. I reached Vellore, a cheerful and relatively clean market town, after a hectic 120-kilometre bus ride from Madras. The CMC Hospital, founded on the first day of this century, is Vellore's pride and joy, and people from all over India come there for treatment. During lunch in the hospital's spartan cafeteria, Jacob John was definite. There were better explanations than plague for the outbreak of disease, both in Beed and Surat. In Beed, the outbreak might have been tularaemia, which can mimic the swollen lymph nodes of plague without its mortality. And the outbreak in Surat might have been due to *Pseudomonas pseudomallei*, which can produce symptoms like those of pneumonic plague – but again without plague's severity. Why, he asked me, had the bacteria not been successfully cultured? The plague bacilli are easy to grow, and the victims, particularly those with pneumonic plague, should have been filled with them.

John told me how, at a recent meeting of the Indian Association of Microbiologists in Poona, he had been told by the Health Minister of the Maharashtra state government (where the Beed outbreak had occurred) not to present his alternative explanations. After a furious argument, he and his co-workers had finally been allowed to give their talks that evening.

I was bemused by this, partly because I was still suffering from a kind of India-shock. There is a dramatic contrast between the rarefied world of science and the raw reality of India. While I was giving a talk at the hospital my wife, chatting with some people in an anteroom, was startled to see a sweeper casually kill a wild bandicoot rat that had strayed in from some nearby ricefield. These rats can be alternative hosts to a great variety of human diseases, including plague. They are everywhere in southern India, often living in close association with humans. When I mentioned this incident to John, he agreed with me that there was no obvious reason why the plague *should not* have happened in Beed and Surat. Conditions were certainly ripe for it. The question was, *did it happen?*

My next encounter with a bit of the plague puzzle came when I flew north to Calcutta, Rudyard Kipling's 'city of dreadful night'.

The impact of this most poverty-stricken of India's cities can be overwhelming. A visitor flying in at night descends over huge areas on Calcutta's outskirts where the only light comes from flickering fires. They mark the areas where hundreds of thousands of refugees from Bangladesh and their descendants still eke out a miserable existence in decades-old shantytowns. On the narrow streets of the city itself, lined with rotting buildings that look like decaying fangs, seventy thousand rickshaw pullers ply this most demeaning and killing of trades amidst stunning juxtapositions of wealth and poverty. A few steps from the door of Calcutta's finest hotel, I saw the motionless body of a beggar child, collapsed from exhaustion, covered with coins tossed by passers-by who must have been seized by a pang of belated sympathy.

Not all of Calcutta, in spite of its vast teeming slums, is a horror story. Residents of the city can wander the superb botanical gardens or take a ride on the new underground system, and the intellectual life of Calcutta is the most exciting in India. But the economy is still reeling, after decades, from the deluge of refugees who descended on it after Partition, and from the collapse of the jute industry. A further blow has been dealt the city by the consequences of a distant ecological disaster. Calcutta's port, on which it depends utterly, is silting up because of the immense quantities of topsoil that are being carried down by the Ganges and its tributaries from the deforested slopes of the Himalayas.

One morning, I picked my way through courtyards covered in rubble and past the decaying stumps of Doric columns, on the way to the building that houses the Calcutta School of Tropical Medicine. The devastation was unusual even for Calcutta. I entered the building through a set of colonnades, webbed with immense cracks, that were

FIG. 5-1 A Calcutta rickshaw puller

precariously supported by a wooden scaffolding. The deputy director, A. K. Hati, awaited me in an office that had been virtually stripped of furnishings. He explained that the newest branch of Calcutta's underground system lay directly below us, and because the excavation had been inadequately shored up his building was collapsing. The entire school was being evacuated and moved to new quarters (which had not yet quite materialized). As a result science had almost ground to a halt.

Calcutta, like every other Indian city, had of course been panicked by the Surat outbreak. Dr Hati told me that his school had tested the blood of 100 of the evacuees arriving in West Bengal from Surat, and that none had shown any sign of plague. He felt confident that the disease had not reached Calcutta.

In spite of its swarming rats, the city has seen no more than a handful of possible plague cases since the 1960s. Hati suggested that this absence of plague might have been an accidental side-effect of the national malaria eradication programme, which had used insecticides so liberally that the rat flea populations had been reduced.

Later, K. K. Mukherjee, another specialist in insect-borne diseases, put the city's obvious misery into perspective. No matter how bad Calcutta might seem, he said, there are worse places. The day before, I had visited Mother Teresa's orphanage on the Lower Circular Road and had seen a number of children who had been stricken with polio. But he told me that the disease is rapidly becoming a thing of the past – coverage with the live polio vaccine in Calcutta is actually now better than in London or New York. Calcutta's advances had been remarkable. A delegation had recently arrived in the city from poverty-stricken north-eastern Brazil – in order to find out how Calcutta had managed to reduce its rate of infant mortality so dramatically in recent years!

Calcutta takes some getting used to, but by the end of my visit I could imagine (barely) how it might be possible to love the place. And the people I talked to agree that, like the rest of India, it is getting more bearable for its inhabitants. The dreadful extremes of poverty and disease of a generation ago that were recounted in Dominique Lapierre's famous *City of Joy* are gradually being alleviated. None the less, I felt a certain relief when I departed for Delhi.

In spite of the inevitable festering mounds of garbage, Delhi seemed positively antiseptic after Calcutta. There I got a very different story about the plague. At the green and pleasant campus of the Lady Hardinge Medical College, I talked with K. B. Sharma, a regional adviser for the World Health Organization. He told me why the first attempts to isolate the plague bacillus from the Beed cases might have failed. The culture plates had only been incubated for one day, and the bacteria grow fairly slowly, so the colonies might have been too small to see with the naked eye. He and his co-workers had subsequently managed to isolate the bacillus from plates that had been incubated for some time, even though they were crowded with the colonies of many other bacteria and fungi. He was not happy with his results, since the isolation had required repeated subculturing, but he was sure the disease had been plague.

Geeta Mehta, the head of the Lady Hardinge Department of Microbiology, was as fascinated by this emerging detective story as I was. She drove me to the north of the city to visit the Infectious Disease Hospital, where I obtained a glimpse of what things must have been like during the height of the epidemic. In late September this small, 100-bed hospital had been overwhelmed by more than a thousand patients. Whenever plague had been suspected, sick people had been sent there from every other hospital in the area.

Dr R. C. Panda, a tall and imposing man, told me that he had worked frantically, often around the clock, for twenty days, and had not gone home for a week during the height of the epidemic. Most of the patients had a high fever, severe cough, acute chest pain, and blood in the sputum. Sixty-six of them were later shown to have antibodies to the plague bacillus, and Panda saw in their sputum immense numbers of cells with the typical 'bipolar' appearance that gives stained plague bacilli the look of tiny safety pins. (Unfortunately, many other bacilli show bipolar staining as well.) And sometimes, though not often, family members came down with the same disease a few hours after the patients, as would be expected from the highly infectious pneumonic plague. In one case, a child arrived ill at the hospital at ten in the morning and his father was admitted with the same symptoms twelve hours later.

Sometimes, he said, recovery could be astonishingly swift after anti-biotics had been given. On 24 September a patient from Surat who had turned up ill at another Delhi hospital was hastily transferred to Panda's hospital. He suffered from severe chest pain and could not answer questions, only murmur 'I am sinking!' Two hours after Panda had begun to administer streptomycin and tetracycline, he arrived at the patient's bedside to find that he was not in bed but had got up to go to the bathroom. 'You have saved me!' he cried to Panda when he returned.

Panda could not understand all the controversy about whether or not the disease was plague. 'Why did the WHO never come and talk to me?' he asked. 'It was I who saw what was going on!'

Later, David Dennis of the WHO team told me that the WHO group in Delhi had tracked down a number of patients, though none from Panda's hospital. Only one of these patients, who had travelled from Maharashtra state, seemed likely to them to have had plague. What had Panda seen? Could there have been a more severe outbreak in Delhi than had been reported, or were the thousand cases in his hospital simply victims of flu, pneumonia, assorted other fever-generating diseases, and hysteria brought on by fear of plague?

Soon afterwards I talked with K. K. Datta, the Director of the National Institute of Communicable Diseases, at his office near the

Infectious Disease Hospital. Joining us in the office was Dr S. Patanayak, who had been in Surat as the acting director of regional control of diseases at the time of the outbreak. Both were passionate in their insistence that it really was the plague. Datta bombarded me with rhetorical questions. How could such a severe and swift outbreak be due to *Pseudomonas*, as Jacob John had suggested? There was no evidence that meliodosis, the disease that is caused by this species of *Pseudomonas*, acted in such a way! And if the disease was simply viral pneumonia, or (as some had suggested) the result of infection by a hantavirus like the Four Corners virus, then why had so many of the patients responded so quickly to antibiotics that kill only bacteria? Datta admitted that his institute had made an error in not trying harder to culture the bacilli, but excused this by pointing out that nobody at NICD had the expertise any more – dealing with plague in India had become a lost art.

In any case, Datta suggested, the matter was moot. Time had been of the essence, and Indian government agencies had moved with commendable swiftness. His job was to prevent the loss of life, and a minimum number of lives had been lost. In a perfect world, it would have been a good thing to construct a solid epidemiological story, but there had hardly been time for such niceties in the rush of events.

Then Dr Patanayak raised an interesting point. Recent instances of plague in India, he pointed out, were often different from outbreaks in centuries past. In particular, it was not unusual to see sudden clusters of pneumonic plague appear in isolation, resulting from the arrival of one person with bubonic plague from elsewhere. This was the puzzling pattern that had been seen in Surat. In the late 1940s, a few such reports had appeared, particularly during the Calcutta outbreak. And I learned later from David Dennis that an outbreak of plague showing a remarkably similar pattern had taken place in Tanzania in 1990, though a detailed report had never been published. First there had been a number of cases of bubonic plague in some remote villages. This had been followed by a dramatic outbreak of pneumonic plague in a hospital in the capital, Dar es Salaam, in which some doctors and nurses had been infected. So the Indian outbreak, with all its puzzling features, definitely had precedents.

Travelling through other parts of India after I left Delhi, I tried to make sense of this swirl of claims and counterclaims. The gap between the bureaucrats and the scientists was plain, and had overtones of paranoia on both sides. Several scientists had suggested to me an explanation for why the Indian government was so insistent that it was plague. The government would lose face with the electorate if they ever admitted that they might have made a mistake and thrown everyone into a needless panic.

But, I wondered, was the panic needless? People in Surat were certainly dying of *something*, and in large numbers. And the preliminary WHO report had supported the government's contention. On the other hand, the official refusal to countenance alternative explanations was unnerving. So far as the rest of the world was concerned, plague was the worst possible explanation for the outbreak, and by insisting on it the government seemed to be effectively immolating itself on the pyre of world opinion.

The paranoia was not confined to the scientists and bureaucrats. I heard discomfiting stories about the existence of a cabal of doctors in Surat who banded together to insist that the outbreak was anything but plague, though they had no proof of any other cause. The probable explanation I was given for their activities was that the financial community of Surat, with its heavy involvement in diamond cutting and textiles, was terrified of the effects the reports of plague would have on their exports. But my most unnerving encounter took place in Rajasthan, where a rich Hindu woman from Calcutta assured me that the whole thing had been a Muslim plot.

I gathered some more pieces of the puzzle in Bombay, the financial powerhouse of India. The first of them was another interspecies encounter, like my wife's in Vellore. It took place just outside our excellent hotel, not far from the Gateway to India. Across the street from the hotel, late on the night of our arrival, we saw the inevitable mounds of garbage. We also saw numerous rats, swarming insouciantly among the many people who were sleeping on the sidewalk.

The next day, I visited Bombay's Haffkine Institute, housed in an echoing building that dates from the eighteenth century. The Institute itself had been founded in 1896, a result of the Russian microbiologist Waldemar Haffkine's success in treating plague victims at the end of the last century. Haffkine, born in Odessa of Jewish parents, had studied with Elie Metchnikoff of phagocyte fame. In India he had attempted without success to make an effective vaccine against cholera, a task that

FIG. 5-2 Rats need not be associated with plague, as is shown by this picture of a rat-infested ward for women at New York's Bellevue Hospital in 1860.

continues to elude scientists at the present time. But he did manage to develop a vaccine against the plague, using killed plague bacteria, that proved to be more effective than Yersin's.

Haffkine's efforts had been clogged with controversy from the beginning. Of the thousands of people who were given the vaccine, nineteen in a remote village died dreadfully of tetanus, apparently the result of contamination by a local health worker. This accident (which eerily foreshadowed the accident decades later that tarnished the early use of the Salk polio vaccine) had effectively destroyed Haffkine's career.

The current director, Vishwanath Yemul, told me that the institute had once had an effective plague surveillance team, which they had managed to maintain for many years. But the apparent disappearance of the plague and cutbacks in government funding had forced the dismantling of the team in the late 1960s. None the less, a good deal of expertise remained to handle the new emergency. Yemul's group had obtained many samples from both the Beed area and Surat, and had even received a number of live rats from Beed. They also obtained samples directly from people who had fled to Bombay from Surat and had then become sick. In every case, repeated attempts to culture the plague bacillus had failed. Most of the time they simply obtained plate after plate of pneumococci.

They were also able to carry out a sensitive immunological test, involving sheep red blood cells, on the samples. In three out of 63 suspected plague samples, they obtained what appeared to be a signal that these people had been exposed to the plague bacillus. But they also examined eighteen people who were either healthy or were suffering from unrelated diseases, and found to their surprise that half of them had signals that were as high as those of the three suspects. It is unusual to find such a large number of 'false positives'. These people must have been exposed at some point to bacteria that resemble the plague bacillus very closely, bacteria that are simply a normal part of India's teeming microbiological life.

Yemul showed me a paper, in press in the *Indian Journal of Medical Research*, that recounted the results of an even more sensitive test carried out on the samples from Surat. This test employs PCR to pick up traces of *Y. pestis* DNA, multiplying these traces to the point where they can be examined easily. The test worked beautifully on known, highly diluted samples of the bacilli, but was completely negative on all the Surat samples. Could the Surat plague bacilli have changed so much that the test did not detect them? He didn't think so – the original strain isolated by Haffkine a century earlier was still being used by a US company to make an effective vaccine, so little seems to have happened to that strain in a hundred years. So far as Yemul was concerned, there was no evidence of plague whatsoever in any of the samples his Institute had seen.

As I left India, I was still scratching my head over this bewildering set of clues. More bits of the puzzle may eventually fall into place, though whether they will ever resolve the mystery completely is moot. Two commissions were established, one by the Indian Academy of Sciences and the other by the national government. Both are headed by V. Ramalingaswami, the former Director General of the Indian Council of Medical Research. All the available data are being re-examined. Confusingly, at the time of writing, their preliminary results from the Surat samples using PCR have given positive results, which is the opposite of what was found by the scientists at the Haffkine

Institute. But the commissions have not been able to find any evidence of plague in the Beed district. For an unexplained reason, perhaps connected with a fear of damaging Surat's economy, the report in the *Indian Express* for 7 February 1995 played down the positive results from Surat while trumpeting the lack of any results from Beed. And more anecdotal evidence is emerging about the Beed outbreak. Infected children with grossly swollen lymph nodes had been observed playing in the streets. Whatever these children had, it was certainly not plague. The plague bacillus continues to play peekaboo with the scientists who are searching for it across India's confusing epidemiological and political landscape.

On my way home, I flew across the Indian subcontinent for one last time. It was the middle of the dry season, and the wrinkled lands of the Deccan were completely obscured by a thick brown haze that stretched all the way to the horizon, a mixture of smoke from cowdung fires, industrial effluents and diesel exhaust. An answer to the question of whether plague had happened in India seemed as far away as ever, though I suspected that Jacob John had been right about the Beed outbreak. But as I stared out the window at India's shrouded landscape one thing seemed quite certain. Only the Band-aids that have been provided by twentieth-century science are preventing an explosion of disease in the overcrowded subcontinent. One of these Band-aids is the widespread and sometimes indiscriminate use of antibiotics. But while Surat and the Beed district had been flooded within days of the outbreak by ten million antibiotic tablets, once the immediate crisis was past there was no attempt to deal with the underlying problem.

The problems threatening India's public health are immense. Delhi is typical – every day, 5000 metric tonnes of garbage are deposited in the streets and alleys of the city. The city's ageing fleet of trucks and refuse compactors cannot take it all away, so that each day hundreds of tons accumulate. There are 66 large refuse compactors in the city's fleet, each capable of handling eight tonnes. Unfortunately, at the beginning of October 1994, only two of these were actually operating. Residents constantly complain of nearby huge dumping grounds that have not been cleared in living memory.

After the apparent plague outbreak, a parliamentary commission was established to look into the question of garbage removal. In spite of such feeble stirrings of the bureaucracy, it became obvious to me on

26 January that the situation had not improved. This was the date of the annual Republic Day parade in Delhi, with its fabulous displays of camels, elephants and brilliantly costumed dancers. A decade earlier, fly-pasts by the air fleet during the parade had been discontinued, because of a near-accident when a plane had collided with one of the many birds attracted by the city's heaps of refuse. For the 1995 parade, in honour of Nelson Mandela's visit, the fly-past was reinstituted, but at the relatively safe altitude of 3000 feet. The birds continued to circle undisturbed above the refuse heaps. Another hasty Band-aid had been applied to one of India's multitude of problems.

The scramble to survive in India is almost as fierce as ever, and niceties such as garbage collection tend to be lost in the scuffle. Yet ironically, the country's burgeoning population has never been better off or better fed. The widespread starvation of past generations has been replaced by relative abundance. Food is everywhere, and India now exports substantial amounts. Of course far too many Indians still go to bed hungry every night. Half the rural population still shows signs of malnutrition, but most of this is due to a lack of balance in the diet, not a shortage of food.

The last of India's wild lands are disappearing, and the filthy air is rapidly dissolving the remnants of five thousand years of civilization. But the country is surviving, and in some ways thriving. Is this surface prosperity no more than the hectic flush on the cheeks of a moribund patient? How long can these stopgap measures be continued? Can they buy enough time for education and the empowerment of women to begin to have an effect? The outbreaks in Beed and Surat may not have been plague after all, but even if they were not their import is not lessened. They are unmistakable signals of how close India is to a public health disaster.

India, of course, is an extreme case. Even more than the rest of the world, it is a kaleidoscope of contradictions and extremes. But will the situation faced by the rest of the world be so very different from India's as we enter the next century? What are the perils from disease that we confront as a species?

PART THREE

Naïve and Cunning Diseases

6

Cholera, the Black One

> ... at an early period, the date of which cannot be
> ascertained, a female while wandering about in the
> woods met with a large stone, the symbol of the god-
> dess of cholera ... The fame of the goddess spread,
> and people flocked from all parts of the country to
> pray at her shrine in Calcutta.
>
> N. Charles Macnamara, *A History of Asiatic Cholera*, 1875.

The temple of Kalighat in Calcutta is devoted to the noisy, furious
worship of Kali, the Black One, the Hindu goddess of death, disease
and destruction. Black goats are sacrificed in a small enclosure every
morning. In the centre of the temple is a shrine with a small black
image of the goddess, and worshippers push and shove each other for
a glimpse. I could not help but wonder, as I was jostled by the throng,
whether there is any connection between the worship of this ancient
goddess and the disease of cholera. Victims of cholera often turn black
as the blood congeals and the skin collapses. It is a story that cries out
for detective work, for if Kali is the goddess of cholera then this disease
must have wreaked destruction among the deltas of the Ganges for far
longer than the historians of the West have supposed.

Cholera is terrifying, but it is also naïve. We know that it has only
recently crossed species boundaries to inflict damage in human populations – though exactly how recently cannot be determined. Since its
first recorded outbreak in 1817, which began in the river deltas of
Bengal near Calcutta, it has swept across the world in seven successive
pandemics. The first six were probably all due to the 'classical' cholera
bacillus, type o1. The seventh, beginning in the 1960s, was due to a

slightly different strain called El Tor. It is this strain that in 1991 leaped its way to Peru and then to almost all the countries of Central and South America. And there may soon be an eighth pandemic, sparked by a new strain called 0139 Bengal that was first found in Vellore in 1992 and which has already caused more than 100,000 cases and more than a thousand deaths in eastern India and Bangladesh.

Mary Jesudason, the microbiologist working in Vellore who found the new strain of cholera, told me that she is uncertain what to make of it. It seems to be no more and no less dangerous than the old 01 strain, and indeed differs very little from it in immunological tests, and yet at the same time it is spreading rapidly through 01's old haunts. It is not, however, replacing the 01 strain completely, and she suspects that 0139 has managed to discover and occupy some new ecological niche.

Water, water everywhere, nor any drop to drink

Cholera epidemics can take many forms. In Peru's ugly, sprawling capital city of Lima, I talked with a man who had anticipated the 1991 Peruvian El Tor cholera epidemic, and who had actually been responsible for blunting its force. Uriel Garcia-Caceres, a clinical pathologist, was born in Cuzco high in the Andes and went to medical school in Lima. Later he did his residency at Cincinnati General Hospital in the US, and on his return to Peru established a strong research programme in tropical diseases that was funded by the US National Institutes of Health.

Ten years before I talked to him, Uriel Garcia had been appointed Minister of Health in the government of Fernando Belaunde that had taken over after a dozen years of military rule. Then, when Alberto Fujimori swept the inept Belaunde from office in 1990, Uriel Garcia left government to return to private practice. Now he heads a private clinic in the centre of Lima, and we met in his cluttered office on the top floor of the modern clinic building.

Uriel Garcia had assumed office six years before the outbreak of cholera in 1991. But he was very well aware of the devastating impact that many other diarrhoeal diseases have had among both young and old, particularly in the sprawling shantytowns that surround virtually every Peruvian city. A scholar of South American medical history, he had read a great deal about the devastating outbreaks of cholera in Ecuador and Chile that had occurred periodically, beginning in 1868, throughout the latter part of the nineteenth century. Peru had only escaped similar outbreaks because of strict quarantine regulations and because coastal cities have not been connected to each other by roads until quite recently. (He remembers that when he was a child in the 1920s it was still necessary for him to travel from Cuzco to the coast and then take a boat north in order to get to Lima.)

Unclean water was the primary agent in the spread of these diseases, and as soon as he took office he began to receive distressing reports about the state of the water systems in the exploding population centres along the coast. Many urban areas in Peru rely on astonishingly primitive water supplies. In Chimbote, 350 kilometres to the north of Lima, there is a permanent river, which is unusual on Peru's arid coast. The town has no formal water system at all. It straddles the river like a town in medieval Europe, and dumps raw sewage directly into it. Even the hospitals of Chimbote take untreated water straight from the river.

Trujillo, 150 kilometres further north, is Peru's second largest city, with a water supply that is a little more sophisticated. But the 61 wells from which it is drawn are connected by a maze of legal and illegal pipes that allows contamination to spread easily. Water is also often stored for long periods in containers in the houses, and faecal contamination can easily build up as people dip it out with their hands.

The capital city of Lima has a far better water system, the direct result of those great outbreaks of cholera in nearby countries during the last half of the nineteenth century. Built by an English company during the first two decades of this century, it was a model system for the time. But Lima's population in 1920 was only 230,000 and is now more than seven million. Little has been done to modernize the system, and by this time it is of course hopelessly inadequate. Chlorination, that great and essential advance in the treatment of urban water supplies, is intermittent at best.

After the epidemic, bureaucrats in Lima claimed that their failure to chlorinate the water properly was the result of a report they had received from the US Environmental Protection Agency that connected chlorination to a small risk of cancer. This seems merely to have been a face-saving excuse to cover up their laxity – while the city of Lima seems to have had the resources to chlorinate its water, the bureaucrats in charge chose not to make the effort. But in towns that were far from Lima, the situation was different. In Trujillo and in many other cities and towns along the coast there was simply no money or equipment for chlorination.

Confronted with this ticking time bomb, Uriel Garcia did what he could. He was aware of the work that had been begun in India and Bangladesh with rehydration therapy, a simple treatment that uses salts and glucose to restore water and essential electrolytes to the body during severe bouts of diarrhoea. Indeed, when he first arrived at the Ministry of Health, he found that a small pilot programme was actually operating.

He swiftly cut through bureaucratic red tape and ordered the pro-duction of millions of small packets of salts and glucose, at a cost of about seven cents each. He knew that it was essential to advertise the packets widely, and his publicity department chose a catchy and rather daring name for them: *bolsitas salvadoras*, little packets of salvation. With a name like that, it turned out to be easy to get across to people the idea that they should dissolve the contents of these packets in the water that they gave to children and other family members who were suffering from diarrhoea. In the first year, during which the programme was put in place in Lima and Trujillo, Uriel Garcia was able to cut the number of deaths due to diarrhoeal illness by forty per cent.

It was very lucky that this stopgap measure was already operating when cholera hit. The first cases of cholera were reported in Chimbote, but the disease broke out almost simultaneously at many places along the coast and soon moved inland. The World Health Organization immediately jumped to the conclusion that the outbreak had been caused by contamination of seafood and vegetables, though all the subsequent studies have shown that in fact it was contamination of the water supplies that was the culprit. These findings came too late for the Peruvian seafood and farming industries, which lost hundreds of millions of dollars as panicked overseas buyers abandoned Peru for other sources.

When the epidemic had run its course there had been more than 400,000 cases in Peru, with over 4000 deaths. This one per cent death rate is actually low compared with other recent cholera epidemics. As recently as 1975, for example, an outbreak in West Africa claimed fifteen per cent of those stricken. Uriel Garcia's little packets of salvation had undoubtedly helped to save many lives during the Peruvan epidemic. In Trujillo, half the patients admitted to hospital had already been treated with the rehydration salts before they arrived.

Were his packets the answer? Uriel Garcia was anything but sanguine. Not only had he been attacked by opposition parties for wasting money on a foolish programme, and by the Church for giving the packets a sacrilegious name, but he later realized that both local and national politicians had used the success of his little packets of salvation to put off doing something about Peru's decrepit or nonexistent water systems. His efforts had been used as a Band-aid, nothing more, and might indeed have left the country worse off than it had been.

The dilapidated water systems, such as they are, continue to creak away. When I turned on the tap in my hotel room in Lima that evening, the unmistakable smell of sewage filled the air.

The Broad Street pump

The water that gushes forth when we First Worlders turn on a tap is usually crystal clear. It can be used for drinking or cooking immediately, but if we incautiously put it in a goldfish bowl the fish will die. This is because tiny amounts of chlorine compounds have been added to kill micro-organisms. When a container of this water is set aside for a few hours, these compounds break down, allowing the water to be poured into the bowl of fish without untoward consequences. The compounds do not, unfortunately, simply disappear. They produce tiny quantities of carcinogens, which are suspected of causing as many as 10,000 cases of rectal and colon cancer annually, but only the lazy bureaucrats of Lima seem to have used this as an excuse to justify discontinuing the treatment. The occasional case of cancer is bad, but the potential consequences of removing chlorine from the water are much, much worse.

As urban populations get larger and the mix of bacteria that can potentially contaminate water supplies changes, it will get progressively more difficult to remove them all before the water reaches your glass.

In many parts of the US, the water has also been filtered to remove larger organisms that can be hundreds of times more resistant to chlorinated compounds than bacteria. Every once in a while, a breakdown of this purification system will allow a pathogen to get through, with dramatic consequences. In Milwaukee in the spring of 1993, large numbers of a tiny protozoan parasite called *Cryptosporidium* escaped into the city water supply because heavy rains had washed a great deal of soil into Lake Michigan. The filtration system through which the city's drinking water was passed became blocked with suspended matter. The parasites probably came from domesticated animal waste that had been carried in with the soil. The prevalence of such waste is a new situation for Lake Michigan, the shores of which used to be heavily forested but are now primarily farmland. The result was 400,000 cases of severe diarrhoea, nausea, fever and stomach cramps, leading to at least six deaths.

Drinking a glass of water containing just a few of these parasites is enough to establish an infection, but most of us can rid our systems of *Cryptosporidium* quite quickly. People with AIDS are not so lucky. *Cryptosporidium* is one of the hundreds of opportunistic infections that can establish themselves permanently in the bodies of AIDS sufferers. And it is an extremely important pathogen in the Third World, where infections are continual – it infects ninety per cent of the children in poverty-stricken north-eastern Brazil.

Water in the Western world was once far less trustworthy. A century and a half ago, the bewhiskered directors of companies that provided water to major European cities were also, in their pursuit of profit, purveying sudden death. The first careful investigation of the connection between water and disease was carried out by one of the great pioneers of epidemiology, Dr John Snow. He doggedly traced the link between the water that was drunk by the majority of the people of London and the repeated outbreaks of cholera that had afflicted that city and other cities of the British Isles from 1832 to 1854.

Snow was a distinguished doctor who had acquired his reputation, not by tracking down infections, but in the new science of administering anaesthetics. Since its first successful introduction in the 1840s in America, anaesthesia (given its name by Oliver Wendell Holmes) had taken the medical world by storm. Snow, one of the first to introduce the practice to Britain, administered chloroform to Queen Victoria

FIG. 6-1 John Snow

during the births of two of her nine children. The Queen's ready acceptance of the new technique helped to defuse the objections of churchmen who claimed that the pain of childbirth had been ordained in the Bible.

Snow's reputation was secure, and he could have continued indefinitely in his comfortable practice, but as a caring physician he was struck by the horrifying aspects of the cholera epidemics, which in 1832 had made the leap across the North Sea and appeared in England's western seaports.

Other diseases have been responsible for far higher death tolls but have far less spectacularly ghastly effects. The cholera victim may die within hours of the first sudden onset of violent vomiting and diarrhoea. The stools are liquid, and once the bowels have been emptied of all faecal matter the victim continues to pass large quantities of watery material. These 'rice water stools' consist of water and cells torn from the lining of the gut, along with swarms of bacteria.

As this violent dehydration continues, the victim's face seems literally to collapse, and great dark hollows form around the eyes. Violent convulsions of the leg and stomach muscles cause agonizing pain. Surgeons who tried to apply the medicine of the day and bleed their patients found to their horror that the blood had been transformed into an almost unrecognizable thick dark liquid.

In Snow's time, about half the victims of cholera died. The remainder recovered with difficulty, after suffering through a period of high fever. So rapid was the onset of the disease, however, that in most cases the people who died did not have time to develop any fever. Their skins became clammy and flaccid, losing all elasticity and becoming corpse-like even before death. And, to add to the horror of the disease, the corpses of the victims would sometimes go on convulsing after death, so that they threw off their winding-sheets.

Some of the doctors of Snow's time realized that death was caused by the body's sudden, violent dehydration. In 1849, the ingenious and farsighted physician Alfred Baring Garrod analysed the blood of victims and found that about a fifth of its volume had been lost, which explained why it had been concentrated to a thick syrupy consistency. However, giving liquids of various kinds to the victims did little good – those who could still manage to drink soon lost the fluid by vomiting.

Crude attempts were made at transfusion. The first had been carried out by a Russian doctor as early as 1830. He injected a small amount of a weak salt solution into the veins of a patient in the last stages of cholera. The patient seemed to rally briefly, but then died. Other such attempts were equally frustrating, sometimes bringing about a brief and tantalizing improvement but almost always followed by rapid decline and death.

Snow considered the disease with a detective's eye for clues. He was struck by the fact that cholera tended to break out in the late summer, at a time when people driven by thirst would drink quantities of water without first boiling it in order to make tea. And while the disease was often passed from one family member to another, its first appearance in a town or city was often quite unexpected. It tended to break out in highly localized areas, particularly poorer ones. He began to collect reports that detailed the quality of the water supply in these areas. It became obvious that people were often drinking water from creeks and streams, and that there had been no attempt to prevent the contamination of these surface waters by raw sewage.

The poverty of the victims was also an obvious factor: 'Amongst the poor ... there is often very little cleanliness ... [W]e find that, when typhoid fever or cholera enters such a dwelling, it is very apt to go through the house. But ... in cleanly families, where the nursing, the cooking, the sleeping, and the eating go on in separate apartments, it is hardly ever found to spread.' Cholera was an involuntary experiment on a very large scale, and Snow realized that it provided him with rich opportunities for understanding the disease. He strongly suspected that some microscopic organism in the water was responsible, but he had of course no proof. He had to proceed by inference, which grew stronger as the number of cholera cases increased and he could start to make connections.

The biggest involuntary experiment of all took place in London, where Snow had his practice. Amidst the frightful toll of the disease during the late summers of 1849 and 1854, he began to home in on the quality and the source of the city's drinking water.

A number of independent water companies some laying their pipes next to each other under the streets, supplied the city. While this water was laid directly into the kitchens of the houses of the upper classes, for the most part it was made available through common pumps or cisterns in streets or squares. The water was usually drawn straight from the Thames, but depending on the rates that were paid by owners or landlords of the properties being served, the companies might make feeble attempts to render it less unappealing.

In years of plentiful rainfall, the flow of the Thames was sufficient to carry out to sea the waste from the hundreds of sewers that emptied into it. But in dry years the flow was slowed dramatically and actually sometimes reversed by the tide reaching in from the estuary far downstream, so that almost undiluted salt water reached into the heart of the city at Battersea.

It was this fermenting stuff that was drawn from the river by enterprises such as the Southwark and Vauxhall Company, which supplied the areas of Lambeth and Southwark on the south bank. It was also drawn up by the much tonier Chelsea Company, which served the

MONSTER SOUP commonly called THAMES WATER

FIG. 6-2 A cartoonist's impression, from 1827, of the state of London's water supply

upper-class houses of Chelsea and Belgravia. The Chelsea Company filtered its water through sand, while the Southwark and Vauxhall Company did not.

Snow reported:

> Many of the people receiving this latter supply were in the habit of tying a piece of linen or some other fabric over the tap by which the water entered the butt or cistern, and in two hours, as the water came in, about a tablespoon of dirt was collected, all in motion with a variety of water insects, whilst the strained water was far from being clear.

People who drank this foul brew on the south side of the river were, as Snow discovered, twice as likely to contract cholera as those on the north side. But the outbreaks that Snow investigated, both in 1849 and

1854, took place in many parts of the city and involved water from at least eight different companies. None of these companies, it seemed, was innocent, if indeed the water was the source of cholera.

During the early part of September 1854, more than 600 people died in a small area of Soho, centred on a heavily used public water pump in Broad Street. These were for the most part the families of tailors and clerks who worked in the fashionable shops of Bond Street and Regent Street. The course of the disease was particularly violent, and very few people recovered once symptoms began.

Snow at once suspected the Broad Street pump, but by the time he examined the pump's water it seemed to be less repellent than the brown foamy brews he had observed coming from the taps in many other parts of the city – or, indeed, that had come out of the same pump a week or two earlier. For some weeks before the outbreak, he was informed, the water 'smelt like a cesspool and frothed like soap suds'. When this happened, people would operate the pump until the water became clearer. But by the time Snow looked at samples of the first water to emerge from the pump it appeared quite clear, with only a few white flecks floating in it. None the less, if it was kept standing for a day or two, it formed a scum and began to stink.

On 7 September, after the outbreak had begun to subside, Snow managed to persuade the Board of Guardians of the local parish to remove the handle of the pump. Did this dramatic action have an effect on the epidemic? Snow himself admitted that it probably did not, since the number of cases in the area was already diminishing. But it was such a striking action that it focused attention on the desperate need to improve the city's sanitation and water supply. Snow had not closed the circle of proof, but the circumstantial evidence seemed overwhelming. The publicity from his investigations led to the passage of a series of bills, proposed by Benjamin Disraeli and other MPs, which over the objections of landlords forced the overhaul of London's ghastly water and sewage systems. As soon as this was done, cholera rapidly disappeared, never to return.

The incident with the pump handle is the one that everyone remembers. It is still commemorated in the name of the John Snow public house in Broad (now Broadwick) Street. (Of course, the pub is frequented today chiefly by people to whom the very idea of drinking water is anathema.)

In spite of Snow's findings, the third International Sanitary Conference, held in Constantinople in 1866, announced officially that it was air, and not water, that was the medium through which cholera spread. Luckily, this pronouncement had little effect on the efforts that had been undertaken in cities throughout Europe and America to effect a general improvement in water supply and sewage disposal.

There is now no doubt that, as Snow suspected, cholera can be carried by faecal matter that is either transmitted by contact or gets into the water. The bacillus, *Vibrio cholerae*, was first seen in 1854 by Filippo Pacini, but was only proved to be the causative agent in 1883 by Robert Koch. Much has since been discovered about its deadly toxin – a protein that has now been turned by resourceful microbiologists into a tool for probing how the living cell works. But even today argument continues about why the bacillus spreads the way it does, and where it lurks when it is not killing human beings.

The fickle bacillus

Cholera is perhaps the most famous of that large number of diseases, some distressingly little-known, which can kill by violent vomiting and diarrhoea. Cartoonists of the nineteenth century depicted cholera as a death's head, with outstretched black skeletal arms representing the blackened blood beneath the skin of its victims. Its swiftness was commemorated as well – it was often drawn as a death's head or as a giant vulture plucking up its victims with a single sweep of its grim talons.

Vibrio cholerae is one of a group of bacilli with thrashing flagella on their ends. It is curved, making it look like a little comma under the microscope. Robert Koch actually called it the 'comma bacillus'. In liquid cultures it can form a scum on the surface, a scum that writhes with motion that is visible to the naked eye as the bacteria thresh about. It is fascinating to speculate that Snow might have seen this scum on the water that had been drawn from the Broad Street pump and left standing.

Why is the onset of cholera so swift? It is fitting that the first real

FIG. 6-3 A mid-nineteenth-century view of cholera

insight was gained in the city of Calcutta, where cholera was so common that it was endemic rather than epidemic. The Indian scientists S. N. De and P. K. Dutta, working in the Medical College of Calcutta in the late 1950s, developed an experimental system that led them to the first insights into how the bacilli caused their damage. They tied off loops of gut in living rabbits, then injected the loops with fluid from a culture in which *Vibrio* had been grown but from which all the bacteria had been carefully removed. The next day the loops were enormously distended with fluid.

Normally, water and salts can pass freely through the intestinal wall in both directions, with a net inward flow of about 200 cubic centimetres a day. This is just enough to keep the faecal matter in the bowels moist, but not enough to be too much of a drain on the body's resources of liquid. The toxin secreted by the cholera bacillus, however, upsets this gentle process dramatically. Specific parts of this large and complex molecule bind to receptors on the surface of cells. This causes an enormous overproduction of the enzyme adenyl cyclase, which in turn produces large quantities of a molecule called cyclic AMP. This important compound acts as a messenger that can penetrate the interior of the cells, causing them to switch on a number of processes full blast. As a result, chloride and bicarbonate ions are excreted, and the cells are unable to take up sodium ions.

The cells that are most dramatically affected by the toxin are not those that line the gut, but rather the nerve cells that control them. Recent experiments have shown that if these toxin-treated nerve cells are prevented from functioning, water secretion in the gut is unaffected. But if they are allowed to function, they trigger a violent expulsion of water into the gut's interior. The loss of salts also causes the nerves leading to the muscles of the abdomen and the legs to misfire, resulting in convulsions.

Direct injection into the bloodstream of water containing the proper physiological amounts of salts can alleviate the symptoms, as we saw earlier. In the century and a half since 1830, when crude injections had first been tried, intravenous therapy has gradually been improved, to the point where a cure is the rule if the disease has not progressed too far. Indeed, when the victim has fallen unconscious – as was the case with the man in Belén who collapsed in front of Nurse Wilma – it is often the only possible therapy. But even now it is expensive and dangerous and requires trained personnel, who are usually in very short supply during a cholera epidemic.

In such an emergency situation, it is essential to get the salts and water into the body through the mouth. It was not until the 1960s that a way to do this was finally found. The discovery came not in a clinical but in a laboratory setting. Several groups of scientists discovered simultaneously that neither a solution of salts nor a simple solution of the sugar glucose can be absorbed by the lining of the gut. But a mixture of salts and glucose is taken up readily. This is not in retrospect surprising. In real life, after all, one does not usually eat pure salt or pure sugar, but foods that contain a mixture of the two, and t is this kind of mixture that the intestine is adapted to deal with.

In 1964, the first therapy using a solution of both salts and sugar was tried on human cholera patients *in extremis* by R. A. Phillips, working in Dacca in what was then East Pakistan and is now Bangladesh. The patients were so far gone that they died in spite of the therapy, which discouraged him greatly, but he was able to show that they had managed to take up some of the solution.

By 1968, medical workers in Calcutta and Dacca had extended Phillips's results, and found that mixing glucose with the salts in equal proportions did indeed help to get the salts across the intestinal barrier and into the body fluids. This simple therapy can be administered by anyone, and now, in outbreaks like the one in Peru, it has helped to reduce mortality from fifty per cent to less than one per cent. Even in the remotest parts of the planet, cholera is no longer a mysterious and irresistible terror.

There are still many puzzles about the cholera bacillus, however. Sometimes, for no obvious reason, it does not kill. John Snow recounted the tale of one astonishing case in which a nurse, confused and fatigued, accidentally drank some of the 'rice-water' that had been evacuated from the bowels of a cholera victim. This utterly revolting and surely fatal mistake actually turned out to have no consequences – the nurse was unaffected. It was only one of many instances in which there did not seem to be an obvious connection between the organisms and the disease.

Even after Koch's demonstration that cholera was caused by the bacillus, such uncertainties were seized on by a number of bacteriologists, notably Max von Pettenkofer, a professor of hygiene at the

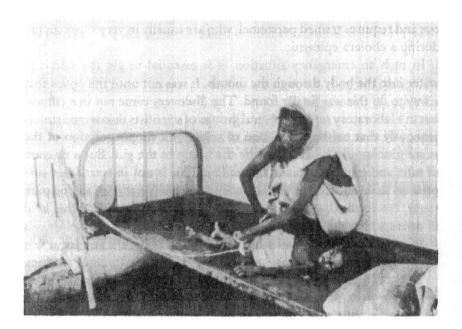

FIG. 6-4 Cholera rehydration therapy

University of Munich. Pettenkofer argued strongly against the drinking water hypothesis. In a famous experiment, he neutralized the acid of his stomach and then drank a culture of *Vibrio*. His stomach swelled a little and he had a mild diarrhoea and nothing else. A faithful co-worker, Rudolf Emmerich, who later succeeded him in the chair at Munich, repeated the experiment with the more drastic results of violent diarrhoea and leg cramps – though Emmerich denied the symptoms until he was forced to admit them during a court case.

While Pettenkofer conceded that *Vibrio* was probably important, he insisted to the end of his life (he committed suicide in 1901 at the age of 82) that it was only one of an unknown number of causes that had to operate in concert.

It is embarrassing for microbiologists to admit it now, but Pettenkofer might have been partly right. Like any other bacterium, *Vibrio* has a hard time surviving in the acid environment of the stomach. In experiments, cholera can be induced most easily in volunteers who have low stomach acidity or who have deliberately neutralized the acid beforehand, giving the bacteria a chance to establish themselves. It is hard to produce a case of cholera by simply ingesting bacteria. Yet Pettenkofer, well aware of the problem, had deliberately neutralized his stomach acid beforehand to give the bacteria every chance. He was certainly justified in thinking that many factors were involved, though his insistence that water supplies were not primarily to blame was monstrously wrong.

Under the whelming tide visit'st the bottom of the monstrous world . . .

As he sampled the water from various parts of London, John Snow was struck by the amount of salt that he found in it. It was easy for him to measure the water's salt content, by using the same chemistry that had made photography possible. He added a few drops of a silver nitrate solution to a sample of water and then exposed it to light – the deeper the brown colour that developed, as the soluble silver nitrate was converted to the insoluble silver chloride, the higher the concentration of chloride. He found that there was far more salt in the water from the Broad Street pump than in water taken from the pipes serving adjacent streets. The different amounts of salt were a function of the various places along the banks of the Thames at which the water companies' pumps were located. The ultimate source of the salt, of course, was sea water from beyond the Thames estuary, brought far up into the river by tidal action.

We now know that it is exactly such mildly salty, nutrient-rich water that provides the conditions *Vibrio cholerae* needs to thrive in nature. Most of the time these bacteria have nothing to do with man. Humans are an incidental host, a brief opportunity to be seized. How the *Vibrio* bacillus manages to seize that opportunity is slowly becoming apparent through investigations of their ecology and molecular biology.

In the early 1960s, *Vibrio cholerae* strains were found in waters of the US Gulf Coast and Chesapeake Bay. Most of these strains seemed at first to be very different from the cholera-producing bacteria – they had different molecules on their surfaces, and most strikingly they did not make cholera toxin. But painstaking molecular detective work by

Rita Colwell and her co-workers at the Maryland Biotechnology Institute has shown that most of the genes of these strains are very similar to those of the strains that produced cholera.

These bacteria are actually well-adjusted inhabitants of the brackish waters of the estuaries in which they live. They can attach themselves to the shells of shrimps and shrimp-like copepods, and are quite at home in the guts of clams, mussels and oysters. Colwell showed that they can also survive long periods of starvation, by shrinking in size and entering a state of dormancy. In a healthy estuary they form an important but not overwhelming part of the bacterial flora. But their numbers increase greatly when the estuary is polluted.

So, of course, do the numbers of a small subset of other species of bacteria and algae. The rich nutrients that flow into a polluted estuary have the effect of simplifying the ecosystem, allowing certain species to multiply at the expense of others. It is not a coincidence that cholera, which was apparently a rare disease in ancient times, started to explode into pandemics as the human population began to grow and sewage from humans and domesticated animals began to run off the land into estuaries in greater and greater quantities. It was in the vast and complex delta of the River Ganges in India that these ecological changes first began to have a dramatic effect.

How were these bacteria transformed from innocent members of an aquatic ecosystem into tiny murderers? Although some strains of *Vibrio* that have been isolated from the US Gulf Coast estuaries are capable of making cholera toxin, most of the *Vibrio* that are found in relatively unpolluted estuaries do not usually carry the gene. In fact, they lack not only the gene itself but an entire section of chromosome on which this gene and the genes for two other toxins are found. There are two possible explanations for this evolutionary change. One is that a killer bacillus has been turned into a harmless one by the loss of this piece of chromosome. The other, and more likely, is that a harmless bacillus has been turned into a killer through the acquisition of this piece of chromosome from elsewhere.

This second possibility was made more likely when it was discovered that this segment of the chromosome has all the characteristics of a *transposon*, a piece of DNA that probably arrived in the bacterium via a plasmid and then became inserted into the bacterial chromosome. We can be near-certain about this because the traces of this genetic surgery are still visible on the chromosome. The sutures of the operation, so to speak, can be seen in the form of two short identical stretches of DNA that flank the inserted region. These are the tell-tale signs of a transposon. The clues are still there, even though the plasmid that was actually involved in transferring the information has not yet been found.

There are further clues in the DNA. While the ultimate origin of the inserted sequence has not been discovered, it is possible to tell that it must have come from an organism very different from *Vibrio* itself.

All DNA, whether it comes from a human, a *Vibrio* or a petunia, has the same general chemical structure. In part it is made up of four compounds called bases, which have been given the short and designations A, T, G and C. The order in which these bases are strung along the backbone of the lengthy DNA molecule encodes the genetic information.

While much of the DNA of a bacterium carries genetic information, the overall composition of the DNA can vary from one bacterium to another. This is because of the so-called pair-rules, which govern the way the bases pair across the two halves of the DNA double helix. An A always pairs with a T and a G always pairs with a C. But there is no requirement that either of the two strands of the DNA should have an equal mix of A, T, G and C. Indeed, at one extreme, a molecule could consist entire of As and Ts (Figure 6-5). At the other extreme, it could consist entirely of Gs and Cs. Both are perfectly good DNA molecules.

In the natural world these extremes are not found. But there is nothing to prevent one DNA molecule from being AT-rich, while another equally functional and information-filled DNA molecule is GC-rich. And indeed, whole genomes can be biased in this way. One of the earliest discoveries of molecular biology was that the DNA of some species of bacteria is heavily biased towards AT pairs, while other species have DNA that is heavily biased towards GC pairs.

So the DNA carries not only the information built into its genes, but evolutionary information as well. The signs of this evolutionary history are subtle, but are as unmistakable to a molecular biologist as the bouquet of a fine burgundy is to a skilled wine-taster. An *arriviste* piece of DNA that has just been inserted into a bacterial chromosome from some other bacterium may be quite different from it in its overall composition, particularly if the insertion event is a recent one. But over long periods of time, the new bit of chromosome will begin to take on the characteristics of the surrounding genetic material. The new DNA will take on the colouration of the DNA that surrounds it, rather like changes in the accent of a person born in England who moves to America.

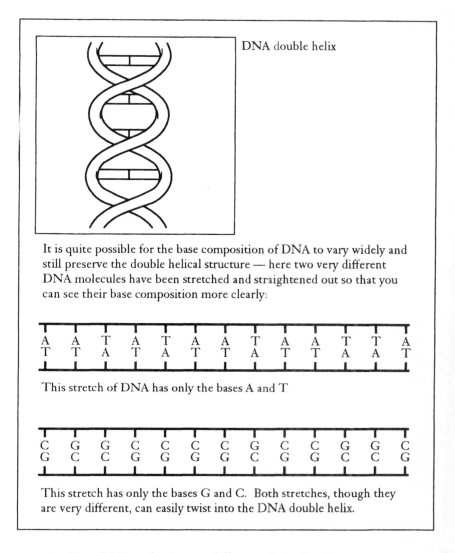

DNA double helix

It is quite possible for the base composition of DNA to vary widely and still preserve the double helical structure — here two very different DNA molecules have been stretched and straightened out so that you can see their base composition more clearly:

A A T A T A A T A A T T A
T T A T A T T A T T A A T

This stretch of DNA has only the bases A and T

C G G C C C C G C C G G C
G C C G G G G C G G C C G

This stretch has only the bases G and C. Both stretches, though they are very different, can easily twist into the DNA double helix.

FIG. 6-5 How DNA molecules can differ greatly in their base composition

A newly inserted piece of DNA sequence can carry other hints about its history. The information coded by the DNA is read in a series of words, in the form of groups of three bases called codons. Each codon specifies an amino acid, one of the building blocks of the proteins that are an essential part of the living cell. The sequence of the codons dictates the sequence of the amino acids. But one amino acid, it turns out, can be specified by more than one codon – for example, the four codons for the amino acid glycine all share the same first two letters, GG, but the third letter can be A, T, G or C. Although the genetic code is a universal language for all life, different organisms can employ different dialects, by using a different mix of these code words.

This difference can be strikingly apparent if similar genes from two very different organisms are compared. We have little proteins called histones, for example, that are essential to forming the structure of our chromosomes. So important are these molecules that they have changed very little in the course of evolution. One of our histones, 104 amino acids long, is identical to the same histone in green plants in all but two of these amino acids, even though it has been a good two billion years since the evolutionary lineages of humans and plants diverged.

During this immense span of time the overall base composition of the two genes has changed, though not very much. The human gene has fifty per cent AT, while the plant gene has sixty per cent. But the codons used in the two genes are very different. Only two out of seventeen codons for the amino acid glycine end with A in the human gene, while nine out of seventeen end with A in the plant gene. Eight out of fifteen codons for the amino acid arginine begin with A in the plant gene, while only one out of fourteen begins with A in the human gene. The codon use of the human gene conforms closely to that of other human genes, while that of the plant gene resembles the genetic 'accent' of other plant genes. If a new gene has become inserted into such a genetic milieu, its overall base composition and its codons gradually begin to resemble the codons of the genes that surround it

This information embedded in the structure of a DNA molecule allows its genetic history to be determined. If the overall base composition and codon usage of a particular piece of DNA conform to the rest of the organism's DNA, this allows molecular biologists to say with some certainty that it has existed in that organism for a long time. But if its composition and codon usage differ markedly from those of the rest of the organism's genes, the likelihood is strong that it has arrived recently from somewhere else. And it is just such a clue that is provided by the piece of chromosome carrying the cholera toxin genes. It has certainly arrived in *Vibrio* from some other bacterium.

As more and more is learned about this collection of genes, and about other genes on the *Vibrio* chromosome that have similarly complex evolutionary histories, their ancestry is becoming clearer. Another bacterium called *Vibrio mimicus*, which can also cause life-threatening diarrhoea in its victims, seems to be the immediate source of some of these toxin genes. But when one tries to follow the trail back beyond *V. mimicus*, it grows colder.

The cluster of genes brought in by the plasmid looks as if it has had a long evolutionary history of its own, during which it has gathered together a variety of dangerous genes into one deadly package. Not only does it carry three kinds of cholera toxin genes, but John Mekalanos of Harvard Medical School has discovered yet another gene in this little stretch of DNA that helps convert *Vibrio* into an agent of sudden death.

Like other bacteria, *Vibrio* are readily killed by stomach acid and are washed out of the intestine – you will remember that Pettenkofer deliberately neutralized his stomach acid before drinking his *Vibrio* cocktail, in order to give the bacteria every chance. But if a few *Vibrio* can survive this acid bath and manage to attach themselves firmly to the inner lining of the intestine, they can avoid being washed away, and within hours will multiply to the numbers needed to produce lethal amounts of toxin. It is exactly this process that Mekalanos's gene facilitates. It codes for a protein on the surface of the bacteria that allows them to cling more firmly to the gut cells.

When one looks at this nasty collection of genes, one almost has the feeling that the transposon has been deliberately designed to confer on *Vibrio* all the properties that it needs to cause havoc in the human gut. The design, of course, was not a conscious one, but was the result of a long period of evolution. Somewhere, presumably among the thousands of bacterial species inhabiting the world's estuaries, there is another bacterium that has evolved this cluster of genes, over a period of tens of millions or hundreds of millions of years, and that has relatively recently donated the cluster to *Vibrio*. Does that bacterium also cause enteric disease, perhaps in some other animal? We will not know until it is found.

When a few toxic strains of *Vibrio cholerae* were first discovered to be lurking in estuaries, it was initially supposed that these strains had a rather complex life cycle. They would infect humans, cause diarrhoea, and be washed into the estuarine waters. Then they would take up residence in marine organisms, and at some point in the future infect more humans. But the alien nature of the transposon that carries the toxin genes makes such a cycle unlikely. Only after *V. cholerae* has picked up the pre-evolved package of toxin genes from somewhere else can it break out from its natural cycle. We know very little about that cycle, except that it does not involve humans

Just as *Vibrio* has its regular cycle, independent of and harmless to humans, the bacterial source of the deadly toxin-carrying transposon must have its own cycle, which may cause harm to other animals but which also does not include humans. The disease of cholera requires the intersection of these two cycles, the transfer of the transposon, and the availability of large numbers of human hosts.

The cycles must have intersected for the first time in the distant past. Just as people hanging around the street corner who are provided with a basketball can form a pick-up team, these sets of highly evolved genes have come together in new combinations and in new organisms, enabling them to take advantage of new ecological opportunities. To judge by anecdotal reports, there were occasional cases of cholera in the cities of ancient Greece and Rome, but no tales of massive cholera outbreaks have come down to us. The most dramatic consequences of the intersection of bacteria and transposons have only developed in the last two or three centuries. This is far too recently for any really large evolutionary changes to have taken place in the genes involved.

Malthus revisited

Oh, Mr Malthus, I agree
In every thing I read with thee!
The world's too full, there is no doubt,
And wants a deal of thinning out . . .
Why should we let precautions so absorb us,
Or trouble shipping with a quarantine –
When if I understand the thing you mean,
We ought to *import* the Cholera Morbus!
Thomas Hood, 'Ode to Mr Malthus', 1832

Most *Vibrio* strains do not carry the gene cluster for the cholera toxin. Those that do carry it tend to live in regions where large numbers of humans have modified estuarine environments. If it had been left alone, the Thames would never have brewed up the deadly combination of brackish water and sewage that gave rise to London's cholera epidemics of the nineteenth century. Such fatal events require manipulation and simplification of the environment by humans. But occasional cases of cholera could and did appear even before rivers such as the Thames had become so extremely modified by human activity.

It is no accident that cholera spreads easily through dense populations, for the more crowded the population of potential victims the more likely they are to contaminate their water supply. And this raises the question of whether cholera has evolved as a mechanism to control population size.

We might be tempted to suspect such a mechanism when we look at the impact of cholera-like diseases on other animals. Avian cholera is a disease of growing importance among both domesticated and wild birds, and it is characterized by the same kind of explosive outbreak that is so typical of the cholera that affects humans. It is caused by a very remote relative of *Vibrio* called *Pasteurella multocida*, and it results in the death of hundreds of thousands of wild birds, chiefly waterfowl, in North America every year. But the disease has a very different set of symptoms from those of human cholera. It is marked by devastating lung infections that produce a copious nasal discharge. Apparently it is a combination of this discharge and the damage to the lungs that kills the birds.*

Large flocks of wildfowl habitually stop during their migrations to feed in the rich tidal ponds near Centerville in the far north of the California coast. They have done so for uncounted generations. But lately these unsuspecting birds face a growing risk of disease as they flutter down on to the shallow brackish ponds. Even in this area, remote from the great sprawling cities to the south, the reduction in wetland habitat has been severe. The result has been that the birds are crowded together in ever greater numbers as they feed. Over the last few years, Shawn Combs and Richard Botzler of Humboldt State University have monitored repeated avian cholera outbreaks.

Ultimately, and unsurprisingly, humans are to blame for these outbreaks. Coastal dunes protect the ponds on the ocean side but there is much pasture on the landward side, heavily grazed by sheep and cattle. Runoff from the manure pollutes the ponds, though they seem pristine to the casual visitor. This increases the productivity of the ponds, which enables them to sustain the crowded bird populations. And it also results in a more favourable environment for bacteria, including *P. multocida.*

Eight species make up the majority of the thousands of birds that visit the area. Some of them habitually dabble in the shallow parts of the ponds and others prefer deeper water. Combs and Botzler found that different species were affected very differently by avian cholera. Generally, the impact was greatest among the species with the largest numbers, and of these the ones that spent most of their time in shallow water were most severely affected. During their study, more than a thousand coots, the most numerous species, died of the disease, while the mortality was far less among the others.

The conditions leading to the spread of avian cholera have obvious parallels with the situation that faced the human populations crowding the festering slums of London a century and a half ago. In northern California a wetland environment has been artificially enriched, in this case by animal wastes, providing an ideal medium for bacteria to grow. There is a dense population of potential victims, which are forced to drink the water from the estuary. And the *P. multocida* bacterium

---

* *Pasteurella multocida* is also responsible for a number of dangerous human infections, including encephalitis. But it is not, at least as yet, responsible for any severe human epidemics.

produces a toxin, although it has very different effects from that of *Vibrio*. The *Pasteurella* toxin binds to the cells that line the birds' lungs, changing their properties drastically. Although its mechanism is different, avian cholera is as fatal as the human kind.

Diseases like avian cholera are telling us something very important about the natural world. Just as we are surrounded by a penumbra of diseases, from which a highly virulent agent occasionally emerges, so other species are surrounded by their own sets of pathogens. Relatively harmless most of the time, some of these pathogens have the capacity to seize a brief window of opportunity when the density of the host population increases. They have done this many times before throughout evolutionary history, and there is no doubt that they will do it again. They are the unconscious agents of population limitation.

Our species is the cause of a double tragedy. It is tragic that our activities are triggering inappropriate outbreaks of these agents of disease in other species, which helps to threaten them with extinction. It is also tragic that when the same kinds of outbreaks happen to our own species, we do not learn from them. We put Band-aids on the problems, allowing them to grow worse as our population continues to expand and our activities dangerously simplify the natural world even further.

Third World countries like India, with the highest rate of population growth, are often those in which disease is most rife. It is a truism among demographers that in our species, diseases do not at the present time act as agents of population control. This is not because diseases *cannot* act in this way, but simply because we are clever enough to be able to ameliorate their effects.

7

A cleverer pathogen

The proximate cause of [typhoid] fever is in all cases
a material and specific agent or virus.

William Budd, *On the Causes of Fevers*, 1839

We should not, for our peace of mind, reflect on the incredible series
of accidents down through the millennia that have led to our own
arrival on the scene. The number of narrow escapes, near misses, and
astounding coincidences that finally result in a particular sperm meeting
a particular egg at a particular time is enough to make anybody reassess
his or her importance in the grand scheme of things. Things could
easily have turned out quite differently. All of us know such stories of
accidental survival in our ancestry. Members of my own family provide
some particularly striking examples, since many of them spent years on
the outer marches of the British Empire, exposed to unusual levels of
risk of all kinds.

My maternal grandfather, a vigorous and adventurous man, spent
the early 1920s as a geological surveyor in India. He travelled widely,
often to unexplored regions. My grandmother, frail but determined,
would accompany him, sometimes being carried in a sedan chair. But
in 1922 she was stricken with severe typhoid fever, from which she
nearly died. It took her six months to recover, and on her doctors'
advice she returned to England.

It was not long before her husband was forced to join her. One day
in the spring of 1924 he was prospecting in the Bokara coalfield west
of Calcutta. He jumped into a dry wash and found a black bear with two
cubs waiting for him at the mouth of a cave. The mother immediately
attacked, and as he turned to flee one of the cubs seized him by the
trouser leg, tripping him up. He defended himself stoutly with his

umbrella and topee, but the mother bear brushed these aside and seized his head in her mouth, shaking him furiously.

Left by the bear and her cubs for dead, he recovered sufficiently to crawl back to his camp a mile away, but his injuries were so severe that his prospecting days were over. He returned to England with little money and no prospects, but managed to call on the remnants of a classical education and become a clergyman – in his case an employment of last resort. The lives of his family were for ever changed.

As a result of some combination of the residual effect of my grandmother's bout of typhoid, the appalling weather that awaited her in England, and the dreadful food that was now all the family could afford, she suffered repeated bouts of illness. Things became really serious in March 1943, when she awoke in the middle of the night with severe abdominal pain. C. C. Holman, the chief surgeon at Northampton Hospital, had no idea what the problem was going to be when he opened her up. He found to his horror that much of her small intestine had become gangrenous. Part of it had adhered to her pelvic bone, causing the whole mass of intestines to twist around and become blocked. He was forced to join the short remaining uninfected piece of the small intestine to the lower colon, simply chopping out the intervening gangrenous mass.

Astonishingly, my grandmother, who was near death during the operation, immediately began to improve. The surgeons were awestruck by what they had done:

> Mr Hankins and I measured the small intestine removed against the side of a three-foot sink, much as a draper measures ribbon. We found there were 19 ft. of small intestine, as well as the caecum, ascending colon and some of the transverse colon.

The operation was so daring for the time that her case was written up in the premier British medical journal, the *Lancet*. To the delight of the surgeon and of her family, my grandmother made an excellent recovery, her health improved, and she lived for another quarter-century.

I have no proof after all this time, but I suspect (and specialists whom I have talked to confirm) that a possible factor that caused the adhesions of her intestine was the typhoid that had nearly killed her two decades earlier. Drilling holes into the body through the intestine is something that the typhoid bacillus does superbly well. The resulting damage can be extensive even among people who, like my grandmother, recover from the disease. Among other things, it can lead to adhesions. These may eventually in turn produce blockage and massive infection by the other swarming bacteria that inhabit the gut.

By a grim coincidence, at the same time as my grandmother was struck down by typhoid in India, my mother came down with the same disease in England. She had been separated from her parents when she was six, as was the custom with English children born in India, and went to live in the house of a severe and tyrannical aunt. She was boarding at a Dickensian girls' school in Derbyshire when at age thirteen she suddenly and inexplicably came down with the disease and lingered on the edge of death for weeks. Remarkably, nobody else in the school was affected.

What is this disease, which affected my family so dramatically? To an even greater extent than cholera, typhoid was in past ages an almost inevitable accompaniment to war and social upheavals. It is caused by a bacillus that is, as we will see, far cleverer than the cholera bacillus. This is because it has had a far longer history of living among humans and our relatives.

Further, we know that typhoid, even more than cholera, is accompanied by a collection of less virulent but still dangerous companions that have evolved along with it and share many of its characteristics. The typhoid bacillus, like the cunning patriarch of one of those villainous families that figure so prominently in Western films, is surrounded by a rogues' gallery of bewhiskered, tobacco-chewing and trigger-happy relatives.

Echoes of a medieval world

Typhus and typhoid were the twin scourges of the medieval world, striking crowded populations that had been made vulnerable by famine or war. Both are marked by dangerously high fever and skin rashes. They are, however, distinct diseases, as the distinguished seventeenth-century Oxford physician and anatomist Thomas Willis was the first to realize. We now know that typhus is spread by lice, fleas or mites and is caused by a tiny organism, somewhere between a bacterium and a virus, called a rickettsia. Typhoid is bacterial, primarily caused by *Salmonella typhi*. The rash that typhoid produces is distinctive but not particularly startling, appearing as a scattering of rose-coloured spots on the chest and abdomen. The stools are often bloody, because of the damage that the bacteria cause to the guts of their victims.

In Europe, typhoid continued as a severe public health problem until near the end of the nineteenth century, even though its manner of spread was already clearly understood. Twenty years before the discovery of the bacillus itself, William Budd, a Bristol physician, realized that it was spread by faecal contamination, though his conclusions (as might be expected) were widely ignored by the conservative medical establishment.

Typhoid could strike anyone at any time, even those who might seem unlikely to come in contact with faecal material. Prince Albert, the husband of Queen Victoria, was one of its most famous victims. At the time of his death in 1861, the disease was killing 50,000 people a year in England. This toll only began to diminish slowly with the passage of various public health acts, which initiated the clean-up of England's urban water supplies.

Interestingly, these minimal efforts had an almost instantaneous effect on cholera – the last English cholera epidemic took place in 1865. But they had a much smaller effect on typhoid, which declined only slowly through the latter part of the nineteenth century. In England there were sixty cases per year among every 100,000 people in 1871, and even by the turn of the century the incidence of the disease had only declined to twenty cases per 100,000.

Why did the disease decline so slowly? Typhoid was relatively difficult to eradicate because the bacillus could persist in the bodies of its victims. When Robert Koch in Germany carefully cultured bacteria from the stools of recovering patients, he found that some of them were still typhoid bacilli. After a few weeks, in most cases, the 'bacteria count' diminished and these patients were no longer walking bacterial time-bombs.

A minority of the patients, however, continued to excrete bacteria, sometimes for the rest of their lives. And these chronic carriers could form foci of subsequent infections. Once their existence had been discovered, they were immediately marked with the dreadful stigma 'unclean!' like medieval lepers. They were forced to live a life apart until well into the present century. The most famous and dramatic of these cases was one of the first to be discovered, that of Typhoid Mary.

Hidden bacilli

In the winter of 1906, Major George Soper of the US Army Sanitary Corps was asked to investigate a puzzling typhoid outbreak. It had happened during the previous summer, at the vacation residence of a New York banker in decidedly upper-class Oyster Bay, Long Island. Six out of the eleven members of the household had been struck down. Typhoid was hardly a normal accompaniment to the summer festivities in Oyster Bay, and the bankers, brokers and lawyers who were thick on the ground there had been thrown into a panic.

Soper carefully ruled out possible contamination of the water and food supplies. The other families in the area had eaten the same food, and the water was very clean. The only unusual factor he could discover was that a new cook had been hired about three weeks before the first typhoid case appeared in the family. The cook had left soon afterwards. Little was known about her except that she was a large, stout and uncommunicative Irishwoman, about forty years old, and had been an excellent cook.

Soper had fought earlier outbreaks of typhoid in upstate New York, and was convinced that the disease could be passed directly from one person to another. He was able to find the cook's name, Mary Mallon, through her employment agency. Information from the agency also enabled him to track down eight families she had worked for during the previous decade. In seven of these families there had been outbreaks of typhoid. In one, virtually everyone but the master of the house and Mary Mallon had been stricken, and the man had actually been so grateful for her nursing help that he had given her $50.

Soper soon found that the trail of infection was continuing. In the brief time since leaving Oyster Bay, Mary Mallon had found employment in a household in Tuxedo, New York, but left after a laundress in the household was taken sick. When Soper finally tracked her down she was working for another family in New York City. With grim predictability, a chambermaid and a daughter of the house's owner had just come down with typhoid. The daughter had died of the disease, the first fatality in the outbreaks that Soper had traced. And when he managed to confront Mary Mallon, he found that she was preparing to leave this house as well.

Soper thought that Mary Mallon must surely have had some inkling of the havoc that she was causing, and would be glad to know that there were commonsense precautions – such as personal cleanliness – that would eliminate the problem. But when he walked into her kitchen to interview her and try to explain the situation, she chased him from it, brandishing a carving fork. She proved equally unco-operative in a second interview.

There was no choice, Soper felt, but to have the New York Department of Health take her into custody. She was tracked down in March 1907 by Sara Josephine Baker, an Assistant to the Commissioner for Health, accompanied by five policemen. Mary hid in an areaway closet behind several filled ashcans, but a sharp-eyed policeman spotted a scrap of calico clinging to the door of the closet. Baker sat on her all the way to the hospital – 'it was like being in a cage with an angry lion'.

Mary Mallon was forcibly kept in various hospitals by the Department of Health for three years, during which time the bacteria count in her stools was continuously monitored. The bacteria came and went. Sometimes they were numerous and sometimes they disappeared for weeks at a time. But there was no doubt that they were multiplying somewhere in her body. Typhoid Mary was, in Soper's words, a 'living culture tube, in which the germs of typhoid multiplied'.

The case of Typhoid Mary received intense publicity. She was confined to separate quarters with only a dog for company. Food would be brought by a nurse, who put it down by the door and then fled. She claimed that she had never been a typhoid carrier, and tried several times to obtain her release by suing for a writ of habeas corpus. Each time judges found that the Department of Health had been justified in its unique and precedent-breaking action.

There was a mixture of reactions to her plight in the newspapers. In some stories she was portrayed as a fiendish villain, in others as a lonely woman unjustly locked away. Even the *New York Times* cautiously took her side. Then the administrative head of New York's Department of Public Health was replaced, and one of the first acts of the new Commissioner was to release her in February 1910. It was explained clearly to her at the time that her release was conditional on the promise that she would not seek employment as a cook.

Three months after her release, Mary Mallon complained to the Board of Health that the city had taken away her only means of livelihood. Soon afterwards she changed her name and disappeared. Soper discovered later that she had worked as a cook in at least two positions during the subsequent five years, and in both houses there had been outbreaks of typhoid. But it was when she obtained a job as a cook in the Sloan Hospital for Women in Manhattan that she could no longer hide her presence. Shortly after she arrived, masquerading as a Mrs Brown, there were twenty-five cases among the nurses at the hospital.

When the outbreak began, the other servants at the hospital jokingly began to refer to her as 'Typhoid Mary', but it was only as the number of cases increased that the hospital administrators began to realize that they might have the real Typhoid Mary on their hands. As suspicion grew, Mary Mallon did what she had done so often in the past – she changed her name again and fled, first to New Jersey and then to Long Island. There, in a house where she had just begun to work as a cook, she was finally captured and subdued after another epic struggle. When the police were refused entry, they were forced to climb a ladder and break into the house through a second-storey window.

This time her incarceration was permanent. Held first at Riverside Hospital and later at a grim remote facility on North Brother Island off the shore of the Bronx, she passed her life in dreadful solitude and finally died of the effects of a stroke in 1938. There were no mourners at her burial, which was carried out hurriedly and secretively.

It seems unlikely that Mary Mallon was incapable of understanding her role as a spreader of disease. She was intelligent and literate, and indeed Soper remarked on the fact that she wrote clear letters in excellent English. Had she simply panicked at the thought of losing her freedom? Were her repeated returns to life as a cook, presumably the only thing she was good at, because of incomprehension of the danger she posed to others, or the result of overwhelming financial need, or both?

Mary Mallon was only the most famous of a sad group of bacteriological pariahs who dragged out a lonely existence in prisons and hospitals throughout much of the first half of this century. Why did these carriers remain perfectly healthy, while still producing the deadly bacilli over periods of years or decades? There had to be a focus of infection lurking somewhere in their intestinal tracts. Once this question was properly posed, it could quickly be answered. The focus turned out, in most cases, to be the gall bladder.

The gall bladder clings to the bottom of the liver, where it collects and stores a greenish-yellow liquid, called gall or bile from its extremely bitter taste, which is manufactured in the liver. Bile is superb at making fats soluble. The gall bladder and the liver are connected through the common bile duct to the duodenum, the first part of the small intestine. When food enters the duodenum, bile pours out of the gall bladder.

We don't really need our gall bladders, since they simply store the bile from the liver until it is needed, and indeed they can be removed without noticeable effect. Animals as diverse as horses and rats have no gall bladders at all, and get along very well without them. There are even some disadvantages to having a gall bladder. It forms, in effect, a little pocket leading off the duodenum. Like the appendix, which forms a similar diversion further down the intestine, the gall bladder is a superb place for bacteria to gather and for infection to begin. And the duodenum is, not coincidentally, the primary site of infection of *Salmonella typhi*.

There are, at the latest count, at least 1500 different strains of *Salmonella*. Most of these are quite harmless, but a few can bring on the devastating symptoms of typhoid. Somehow, the bacteria that first infected Mary Mallon were able to reach her gall bladder and live there for years, periodically sending their progeny down through her common bile duct into her duodenum.

It is possible to reconstruct the series of events that led to the transformation of Mary Mallon into Typhoid Mary. At some point, perhaps when she was quite young, she came down with a case of typhoid fever. (She always claimed that she had never had the illness, which might have been a lie, but of course most people are very uncertain about which childhood diseases they might have had.) We have no idea of the severity of the infection, though it was probably mild since the death rate from the disease in the days before antibiotics was between twenty-five and fifty per cent.

The bacteria themselves probably came from another carrier. Except when typhoid flares up in an epidemic, the majority of cases of typhoid in the temperate zones are caught from carriers like Mary Mallon.

Mary caught typhoid because her defences failed. The first line of defence was her stomach, with its concentrated acid that kills most micro-organisms and destroys their toxic proteins. Perhaps she had drunk a great deal of water that day, so that the acid in her stomach was diluted. Perhaps she had eaten food that had absorbed much of the acid – a few baking soda biscuits might have done the trick. Whatever the cause, some of the bacteria had managed to get through her stomach, past her pyloric sphincter, and into her duodenum. Here the surviving bacteria came up against another formidable array of defences.

We currently live in a world of sterile, refrigerated and processed foods, but our ancestors often dined off rotted meat from carcasses and other highly contaminated food. It is difficult for us to imagine the kinds and varieties of horrific micro-organisms that their gastro-intestinal tracts must have had to deal with. As a result of millions of such confrontations, our digestive tracts have evolved to defend us against a huge range of bacteria, viruses and protozoa, along with many different kinds of toxins.

The inner surface of the small intestine is covered by tiny lymph-node-like clumps of cells called Peyer's patches. Even in a young child, the Peyer's patches are soon exposed to a huge range of invaders, which carry a great variety of proteins and other molecules on their surfaces. Seventy per cent of the antibodies we produce are manufactured by cells in the Peyer's patches. Most of the invaders, clumped and immobilized by these antibodies, are quickly destroyed by other specialized cells in the patches. A few, however, carry proteins or carbohydrates that the Peyer's patches have not seen before. The typhoid bacilli that had managed to penetrate this far down Mary Mallon's gut fell into this category – they were invisible to her immune system.

In fact, they could be more than simply invisible. They were covered with tiny hairlike projections that could recognize some of the cells of the Peyer's patches and bind strongly to them. Ordinarily these cells, called microfold or M cells, would be alerted to these alien bacteria by the presence of antibodies, but this time the bacteria were the masters of the situation. They stimulated the surfaces of the microfold cells, which began to ripple furiously and to engulf the invaders, which is just what they would have done with bacteria that they recognized. But once safely inside, the typhoid bacteria were not destroyed but were able to invade adjacent cells of the Peyer's patches.

The infected Peyer's patches became inflamed, and soon bacteria were released in large numbers into Mary Mallon's bloodstream. Her temperature began to rise as her body finally responded to the invasion. Since Mary Mallon survived, it seems likely that the really unpleasant manifestations of typhoid did not develop fully. In a severe case the fever rises dramatically, delirium ensues, and there is a risk of encephalitis, pneumonia and heart failure. The Peyer's patches begin to die, causing perforations in the wall of the small intestine – the same damage that probably led to the eventual adhesions and tissue destruction suffered by my grandmother. Perforated like a sprinkler hose, the intestinal wall no longer acts as a barrier to bacterial invasion, and massive peritonitis can develop.

Mary Mallon, like the majority of typhoid victims even before antibiotics, managed to avoid the most unpleasant of these consequences and make a full recovery. Most of the typhoid bacilli were flushed from her gut or destroyed by her laggardly but eventually effective immune system. Most, but not all. Some had entered her bloodstream and been carried to her liver, where they had invaded her common bile duct and crept down to the gall bladder. They managed to invade the cells lining the bladder. And there they stayed, keeping a fairly low profile.

Mary Mallon's gall bladder suffered a bit from the bacterial invasion, though not enough to inconvenience her. Its lining was now continually inflamed by the slow multiplication of the bacteria within it. She might have had gallstones (many typhoid carriers do), and these concretions would have impeded the natural flushing action of the bladder and prevented it from healing itself. Slowly, over the years, as she moved from one unsuspecting family to the next, her gall bladder dripped its deadly cargo of typhoid germs. And these germs were of the same type that had invaded her in the first place, invisible to the immune systems of the people they infected in turn.

Unlike Mary Mallon, most chronic typhoid carriers who were tracked down during the early part of this century co-operated with the health authorities. They became resigned to an isolated, carefully sanitized existence. Many ended up in hospitals, living out their days largely in the company of other chronic carriers. And this exile could continue for many years, for even the first generation of antibiotics could not seem to rid these carriers of their tenacious bacilli. As late as the early 1960s, the treatment of choice was a drastic one – excision of the gall bladder itself. In 1961, the physician R. G. Main reported that this operation had cured eight out of nine patients who had been sequestered for years in a hospital in Stirlingshire, north of Glasgow in Scotland. None of these patients had responded to repeated antibiotic treatment.

More powerful antibiotics have now become available, making such drastic intervention unnecessary. The strange medieval exile of the typhoid carrier is now a thing of the past. But carriers are not. They are still responsible for repeated outbreaks in industrialized countries, as well as in many parts of the Third World.

Not everywhere, however. As more pieces of the puzzle fall into place, the picture that emerges with typhoid, as with plague and cholera, is one in which the bacillus has sometimes been forced to behave abnormally in order to survive. Mary Mallon's gall bladder infection was caused by a strain of typhoid that was living at the very edge of what was possible for it. It had been selected to spread under difficult and marginal conditions.

Escaping from the tropics

Although Robert Huckstep, a physician working in Nairobi during the 1940s and 1950s, saw and treated thousands of cases of typhoid, he found that almost none of them were chronic carriers. He was puzzled by this observation, since chronic carriers can make up five per cent of the infected population in northern Europe and are currently the main source of typhoid outbreaks there. Why this difference? The most likely explanation, he thought, is that in countries with warm climates, where contaminated sewage is everywhere, the strains of bacillus that are most successful are those that can re-enter the environment quickly. There is no need for carriers, because the opportunities for renewed outbreaks through sewage contamination are so plentiful. In northern Europe, however, where even in the filth and squalor of medieval times the cold winters must have reduced the bacterial count, this means of spread was more difficult. Strains of bacteria that could colonize the gall bladders of a few of their victims might then gain the advantage – even at the cost of sometimes producing a particularly severe form of the disease, since in order to reach the gall bladder they would have to make a more dramatic invasion of the body.

The typhoid bacilli of temperate zones are living at the very edge of their geographic distribution, and are forced to find refuge in the bodies of some of their hosts. Other strains, common in Africa, can also cause the severe symptoms of typhoid but are not so invasive that they can penetrate to the gall bladder – or if they do, it may be that they are quickly flushed out.

Salmonella typhi, unlike *Vibrio cholerae*, seems to be unique to humans. Indeed, genetic studies show that it has been unique to us for a fairly long time. Measurements of the properties of many different gene products show that it is clearly different from the hundreds of other *Salmonella* strains that have been examined.

It turns out that most of the dozens of strains of this bacterium that can infect humans do not take this route of deep penetration and subsequent concealment. Instead they produce a milder infection of the gastrointestinal tract, resulting in diarrhoea that helps them to spread. These mild cases, lacking the high fever, spots on the skin and severe damage to the lining of the gut, are usually classified as enteric fever rather than the more fearsome typhoid fever. As long as these cases receive a reasonable amount of care and are liberally treated with antibiotics, they are normally not fatal. It is these milder bacteria that form the main pool of variation in the group of *Salmonella* strains that infect humans. But lurking among them are the occasional *S. typhi*, and of course all the pieces of genetic information that are needed to construct virulent strains of this species are always present as well.

Fascinatingly, strains of *S. typhi* have recently been found to fall into two groups, one that is found only in Africa and the other inhabiting both Africa and the rest of the world. Unlike the new strains of cholera, many of which seem to have arisen – or at least to have become important – only over the last few decades, this division seems to have taken place a very long time ago. It is even possible that the globally distributed strain migrated out of Africa at the same time as the spread of our ancestors throughout the Old World, a spread that has been variously suggested to have taken place somewhere between 200,000 and two million years ago.

If this widespread strain of *S. typhi* was the one that accompanied these human migrants, why did the African strain remain behind? Is the global strain the one that can invade the gall bladder, resulting in carriers, and is the African strain unable to do so?

These questions have not been fully answered, not so much because of technical problems as because of the fact that *S. typhi* is a very dangerous organism. The microbiologist Eduard Grosman, now at Washington University in St Louis, normally works with the far less dangerous bacillus *S. typhimurium*. This bacillus kills mice but not humans. A few years ago, while he was a postdoctoral fellow on our campus at San Diego, he began doing experiments with *S. typhi* and immediately came down with a severe case of typhoid fever. He had made the mistake of transferring these deadly bacteria from one culture to another by sucking them up into a pipette using his mouth, something that all protocols for the study of pathogenic bacteria absolutely forbid. He described it to me as the worst illness he had ever had. For several weeks after he was no longer infectious, he could walk no more than a few yards without resting. Now recovered, he has returned to the study of the safer, if epidemiologically less exciting, *S. typhimurium*.

Groisman's experience gives us some idea of just how dangerous *S. typhi* can be. He certainly did not suck up millions of bacteria in the course of his mouth-pipetting – his infection was probably due to a few hundred or a few thousand bacteria that had become suspended in a tiny mist of aerosol droplets within the tube of the pipette.

Why is this strain of *Salmonella* so fearsome? There are three major sources of its virulence. The first is the ability of the bacteria to adhere to the cells lining the guts of their victims, particularly those gourmand M cells. The second is their ability to penetrate the membranes that surround the M cells and other cells of the body, and to survive once they are inside. And the third is their ability to prevent an immune response, and indeed to invade the very cells that would normally manufacture antibodies that could destroy them.

A bewildering variety of genes have been isolated that affect these various properties. Groisman and his collaborators have contributed some of the molecular pieces to this puzzle, with their discovery of one of the clusters of genes that allow the bacteria to penetrate the cells of the intestinal wall. There are two remarkable things about this gene cluster. The first is that a very similar cluster – so similar that the various genes are arranged in exactly the same order along the chromosome – is found in a distant relative of *Salmonella* called *Shigella*, a bacterium responsible for much of the severe dysentery that kills children in the Third World. And the second is a fact that struck Groisman immediately. In both of these bacteria the composition of the DNA that makes up this cluster is unusual, clearly different from the composition of the DNA of the rest of their chromosomes. It is very clear that this cluster has entered these two types of bacteria from some other and very different bacterium that remains to be discovered. There may have been more than one donor bacterium involved, for the entry of this alien DNA appears to be the result of two different and independent genetic events. Both *Salmonella* and *Shigella* have obtained at least some of the genes that make them so deadly from these unknown and mysterious donors.

What and where are the donors that gave their genes to *Salmonella* and *Shigella*? We simply do not know yet – the mystery is as great as the identity of the unknown bacillus that is the source of the genes for cholera that have been passed on to *Vibrio*. We can none the less infer a few things about them. The mystery donors seems unlikely to be as dangerous to us as *Salmonella* and *Shigella*, or else they would have been discovered by this time. But it does seem possible that they are highly dangerous pathogens in some other organisms. The cluster of genes that helps to cause typhoid when it is introduced into *S. typhi* is so lethally dangerous that it must have honed its lethality over and over again in the past.

This gene cluster is not the only threatening set of genes in the universe inhabited by these pathogenic bacteria. Perhaps the most versatile borrower among *Salmonella*'s relatives is *E. coli*, which in one or another of its pathogenic forms contributes to between a quarter and a half of life-threatening diarrhoea in the Third World.* At the latest count *E. coli* has been found to use ten different mechanisms to invade and disturb the metabolism of the cells lining the gut. Most of the genes for these mechanisms are carried on bits of DNA, either on viruses or on plasmids, that various strains of *E. coli* have picked up and lost again in a complex and ever-changing pattern. So numerous and widespread are these strains that their mix is continually altering, allowing the *E. coli* population to rapidly increase or decrease its pathogenicity according to the selective pressures of the moment.

The search for the bacteria that are the ultimate source of all these

---

* *Shigella* is, except for its enhanced pathogenesis, almost indistinguishable from its close relative *E. coli*.

plasmids and viruses, like the search for the mysterious organisms that harbour the original DNA which gave rise to the cholera transposon, will be very instructive. We must learn more about these diarrhoeal diseases, where they came from, what their hidden progenitors are, and what forces them occasionally to become highly dangerous. So far as day-to-day survival in the tropics is concerned, it is primarily these diseases which shape and shorten lives, and which contribute to the sense of hopelessness and fatalism that pervades so much of the Third World.

Cholera and typhoid outbreaks are easily preventable, and even with primitive technology can rapidly be stopped once they begin. They are the eye of the hippopotamus. But it is the body of the hippopotamus that is the real problem, for these diarrhoea-producing organisms have the potential to produce spectacular outbreaks even as they go about their grim daily work. And as we have seen, even the milder strains of the typhoid bacilli can be deadly when combined with other enteric bacteria. Because of typhoid's connection with *Shigella* dysentery, and even with the normally non-pathogenic *E. coli*, it is a far more subtle and complex enemy than cholera. Yet typhoid has had to learn to become even more subtle as it moves away from the tropics. While it may kill immediately even outside the tropics, it can also sometimes persist in recovered patients for years.

Is this a general pattern? Is there something about the survival of diseases in the temperate zones that requires them to be more resourceful? And is this one reason why there are fewer diseases in the temperate zones? This too seems to be an evolutionary pattern, a pattern as striking in its own way as the remarkable fragility of the mutant organisms that give rise to plagues. In the next section we will look at two diseases that have escaped from the tropics to the temperate regions, and discover how difficult it was for them and how vulnerable they have turned out to be as a result.

PART FOUR

The Challenge of the Temperate Zones

8

An ague very violent

Humboldt has observed, that 'under the torrid zone,
the smallest marshes are the most dangerous, being
surrounded, as at Vera Cruz and Carthagena, with an
arid and sandy soil, which raises the temperature of
the ambient air' ... In all unhealthy countries the
greatest risk is run by sleeping on shore. Is this owing
to the state of the body during sleep, or to a greater
abundance of miasma at such times?

Charles Darwin, Voyage of the Beagle, 1845

Throughout most of World War Two, Britain struggled to retain its
Far Eastern possessions at the same time as it fought to keep from
being overwhelmed by the Germans on the home front. Its armies
became stretched to the point of near-invisibility. Even by early 1943
they had been able to do little about the Japanese thrust into the northwest quadrant of Burma, which was coming close to the heavily-jungled
Manipur state on the Indian side of the border. Then there was a
short pause as the Japanese halted to resupply, which gave a small
British-Indian army, hastily assembled under General Slim, the chance
to engage them.

One member of this ragtag army was an uncle of mine by marriage.
He had only just managed to finish a hurried and totally irrelevant
training course in assault landing in Maharashtra before he was thrust
with his fellow trainees into the steaming Burmese jungles. There he
took part in repeated battalion-strength forays against the enemy's
forward positions. The Japanese, exceedingly skilled by that time in
jungle warfare, inflicted heavy casualties. My uncle was one of them.

Just on the other side of the Indian border at a town called Tiddim, in the course of a close and confusing engagement, he was wounded by a mortar-bomb splinter that pierced his left tibia. Although he was able to go on fighting for a few more days, the wound soon became infected and a large jungle sore developed. To his huge relief, for the fighting had been vicious and often hand-to-hand, he was invalided out and sent to Officers Training School at Mhow in west-central India.

The trip took three days, and by the end of it he was very ill. It soon became apparent that the fever brought on by the infected wound had become complicated by something else. The first doctor he saw had cavalierly diagnosed his problem as indigestion, but his symptoms grew in complexity and unpleasantness as time went on. Every day, a fascinated group of bemused doctors formed themselves around his bedside and then retreated to the other end of the room to consult. After much discussion, they agreed that they were confronted with a combination of amoebic dysentery and malaria. No sooner had this diagnosis been reached than my poor uncle came down with pneumonia as well. Whether the depredations wrought by one parasite cancelled out those brought on by the others is unclear, but he remembers surviving with the aid of an early antibiotic treatment called May and Baker Powder No.163.

It is impossible after half a century to sort out the freight of parasites my uncle had picked up in his few weeks in the jungle. He remembers that the fever tended to occur every third day, which would suggest that his overt symptoms were caused by the relatively benign but still nasty parasite *Plasmodium malariae*. But this may not have been all that he acquired. Infection by another parasite, *P. falciparum*, which causes the most dangerous of the malarias, is often accompanied by severe diarrhoea, and this might have fooled the doctors into thinking he had amoebic dysentery. Yet the symptoms seem wrong, for the chills and fevers of this more severe type of malaria occur every second day. It is quite possible that he picked up both, since infections with more than one type of malaria are common.

However, the jungles of Burma and north-east India harbour yet another type of malaria that produces fevers every second day. It is caused by the parasite *P. vivax*, and we can say firmly even after a lapse of half a century that he had certainly acquired this kind of malaria at some point during his time in the jungle. Indeed, *P. vivax* and *P. malariae* are the only ones of the jostling crowd of parasites thronging his bloodstream of which we can now, after this long period of time, be certain.

After his recovery, he had no more attacks of malaria for more than a year. Then, in January 1945, when he and his company were on manoeuvres in tropical heat in the Siwalik foothills of the Himalayas, the temperature suddenly plunged and they were caught in a freezing hailstorm. Two days later he was hit by a violent malarial attack, with chills and fever coming quite clearly this time every two days, and from which he took a long time to recover. This was a true relapse, showing the classic pattern of a *P. vivax* infection. The relapse can come as much as a year or two after the original infection, even after all obvious signs of malarial parasites have disappeared in the interim. He recovered slowly and, it seemed, permanently – after this flare-up the *P. vivax* parasites seem to have departed from his bloodstream for good.

My uncle returned to England in the autumn of 1945, and except for a few brief trips abroad he has remained there ever since. But accompanying him, it turns out, were the parasites of that every-third-day type of malaria, caused by *P. malariae*. These parasites remained undetected for decades, lurking quietly in some as yet undiscovered corner of his system. Then, over a period of six years, starting in 1987, he had three severe flare-ups of the disease, the most recent occurring in June 1993.

What had happened? How had these parasites survived half a century of dreary English winters, far from the jungles and swarming mosquitoes of their homeland? Why had they persisted when all the other roistering parasites that had turned his bloodstream into a kind of Bartholomew Fair back in the hospital at Mhow had not?

Whatever those other parasites might have been, they had long since disappeared from his system. But the *P. vivax*, and later the *P. malariae*, in his body were simply doing what these parasites have been doing from time immemorial – they were *extending their range*, penetrating new and progressively more unfavourable environments and surviving in the mosquitoes and people they found there. How these tiny organisms do this, their resourcefulness in doing so, and the dangers they pose to our species, illustrate a fascinating aspect of the dynamics of parasites. When times are tough for a parasite, it must use all its resources in order to survive and spread. When times are easy, it may lose those abilities – or, of course, it may never have acquired them in the first place.

The malaria parasites that afflict our species illustrate what happens in both hard and easy times. Our growing knowledge of their physiology and molecular biology has enabled us to understand the story in greater detail than ever before, and in particular to understand why some of our battles against malaria have been successful while others have been such disastrous failures. In this chapter I will trace the story of some of these interactions. As with the other plagues we have met, they form a meshwork of clues that, taken together, will lead us to understand why some parasites are so much more ingenious than others, and why the temperate zones are so different from the disease-ridden tropics.

Malaria parasites are much more complicated, and therefore much more genetically flexible, than the relatively stupid bacteria we have met so far. This is because they consist of cells that are very like the cells making up our own bodies. These parasites cause immense damage – two-fifths of the world's population is at risk of contracting malaria, 110 million new cases are diagnosed annually, and the disease directly causes between one and two million deaths a year. Most of the deaths that can be directly attributed to malaria are among children, but there is no doubt that many more deaths of both children and adults are due to a combination of malaria and other diseases. The effect of malaria is so great that it forms a kind of endemic plague, causing in sum far more deaths and misery than most epidemic plagues.

And yet, as we will see, it is vulnerable. There are two reasons for this. First, malaria has only moved away from the tropics with great difficulty, and the further it ventures from the equator the more it must call on all its resources to survive. And second, even in the tropics, human intervention can have a surprising effect on the disease.

The mosquito connection

The Scottish doctor Patrick Manson, working in China between 1871 and 1890, was the first to find a connection between insects and disease, in this case the dreadful swelling known as elephantiasis caused by the blockage of the lymphatic system by roundworms. He found that mosquitoes were able to pick up the tiny immature worms in the blood-stream of an individual suffering from the disease.

Manson did not do the obvious experiment of allowing these mosquitoes to feed on a healthy person, not because of scruples (in those days native people could be coerced into taking part in all sorts of hair-raising experiments) but because he did not understand the mosquitoes' life cycle. He thought that a female mosquito took only one blood meal during its lifetime, so that it must pass its freight of worms on to another human host in some fashion other than by biting them. Perhaps when the mosquitoes died after laying their eggs they released the worms into the water, so that the unsuspecting victims picked them up from the water supply.

Manson's intuition, which had led him to the mosquito vector of the worms, failed him at this point. The mosquitoes were in fact active transmitters of the disease, transferring the worms from one human to another. His incorrect guess about how the life cycle of the elephantiasis parasite was brought to completion would later lead his protégé Ronald Ross badly astray as Ross tried to solve the even more difficult problem of malaria.

Elephantiasis, with a low transmission rate per bite, tends to afflict indigenous populations. Malaria was a disease with far higher visibility, since it cavalierly struck down the representatives of the colonial powers as well as the local inhabitants. By the time Manson returned to England, although the mode of transmission of malaria was still a mystery, the parasite itself had been discovered. It had been found in 1880 by Alphonse Laveran, a French army surgeon stationed in Algiers. Laveran had examined the blood of infected soldiers and saw, along with the usual immense numbers of red corpuscles and the occasional white cell, small crescent-shaped transparent cells that he could not find in the blood of soldiers who were malaria-free. Laveran might easily have missed these cells, except for the fact that they often contained granules of dark pigment. We now know that this pigment consists of the broken-down remnants of the blood protein haemoglobin, on which the parasites had recently been dining.

Laveran also saw a quite inexplicable phenomenon – sometimes, as he observed them, the cells would swell and extrude large numbers of what appeared to be threshing flagella. These creatures living in the blood of his patients were certainly very different from the simple

bacterial cells that cause anthrax and tuberculosis. They were frightfully active, elaborate protozoans, cells with real nuclei and complicated life cycles, carrying out many more elaborate processes in the bodies of their hosts than the simpler bacteria could ever manage.

Laveran's discovery was greeted with ridicule by the medical community, which by this time should not surprise us. And indeed, he was a long way from proving that these tiny threshing creatures were the cause of malaria, since the soldiers stationed in Algeria suffered from many other diseases as well. But his finding set a number of wheels in motion.

In the last part of the nineteenth century Italy was still prime malaria country, and in 1889 two Italian scientists discovered the parasite that caused the deadliest malaria of all, *Plasmodium falciparum*. Then in 1890, Battista Grassi, a professor of anatomy in the malarious town of Catania on the east coast of Sicily, found yet another malarial parasite, *P. vivax* (the same species that infected my uncle in Burma).

Grassi had ample opportunity to study both humans and animals who were affected by the disease. He and the other Italian workers now found themselves in the forefront of malaria research.

Grassi soon made another discovery, the importance of which was not realized until later. Malaria is not confined to humans. The blood of birds, too, swarms with malarial parasites which look very similar to the human parasites, and sometimes they can produce a range of symptoms very similar to those of human malarias. Further, he even suspected that mosquitoes must have something to do with transmission of the disease. This suspicion was fuelled in part by Patrick Manson's inspired guess about the cause of elephantiasis of a few years earlier and in part by Grassi's awareness that a number of writers, both ancient and modern, had suggested that malaria, swamps and biting insects were somehow related.

Experiments seemed to support Grassi's hunch. His collaborator Amico Bignami transferred a tiny drop of blood, equivalent to a mosquito blood meal, from a patient into a healthy volunteer, and was able to induce malaria. But, through bad luck, Grassi and Bignami did not

OPPOSITE:
FIG. 8-1 Alphonse Laveran and the malaria parasite in blood cells

stumble on the group of mosquito species that transmitted the disease. Although the parasites were indeed picked up from infected volunteers by the various species they tried, they did not survive once they had entered the mosquitoes.

Meanwhile, back in England, Manson had met a young India-born doctor, Ronald Ross, whom he quickly converted into an enthusiast about the problem. Manson succeeded in persuading the Colonial Office to send Ross to India, in the hope that he would be able to work on malaria.

Following the flagellum

Ross, the central character in this unfolding drama, is also the most complex. Truculent and egotistical, he managed to alienate many of his senior commanders in India from the moment of his arrival. His important role in the discovery of malaria transmission has been marred by the way he treated his scientific contemporaries, notably the Italians. Yet he also brought enormous enthusiasm to the problem. He worked long hours under appalling conditions, painstakingly dissecting mosquitoes and peering at their tiny remains in the tropical heat through a cracked portable microscope of his own invention.

Ross found that the malaria life cycle was far more complicated than anything that had been tackled by microbiologists before. Again and again, he observed the parasites extruding their flagella in the stomachs of mosquitoes. Then, after this interval of activity, they would frustratingly stop moving and do nothing further. He was hampered by the fact that he knew nothing about the mosquitoes of India, even though Manson had advised him to go to the British Museum before he left England in order to learn to distinguish the various species. His notebooks, and his letters to Manson, are filled with references to grey-bodied and brindled mosquitoes, designations reminiscent more of *The Wind in the Willows* than a scientific publication. And unfortunately, none of these mosquitoes, of the genera *Aedes* and *Culex*, was able to transmit malaria in humans.

None the less, gradually, painfully, he made progress. He found that the process of flagellation could be induced by the drying of the blood when the cells were put on a microscope slide or entered a mosquito's

FIG. 8-2 Ronald Ross and a feeding *Anopheles* mosquito

stomach. He even played with, and temporarily discarded, the idea that the mosquitoes might transmit the parasites through their bites. After a temporary hiatus caused by a bout of the very disease he was studying, he began to concentrate his attention on the large, rare but distinctive mosquitoes that he picturesquely called 'dapple-winged'. They belonged to the genus *Anopheles*.

Here, finally, he began to make progress. Although he was continually distracted from his researches by problems with the military bureaucracy – they seemed determined always to transfer him to parts of India where he could not pursue his malaria research – he managed to snatch a few months for his studies in Secunderabad, the sister city of Hyderabad in Andhra Pradesh. There, investigating *Anopheles* mosquitoes that had recently fed on a malarial patient, he finally found something interesting. A few days after the mosquitoes had fed on the patient, pigmented cysts began to form in their stomach walls, something he had not seen in any other species. These pigmented cysts were only found in *Anopheles* mosquitoes that had fed on malarial blood.

Ross recounted these discoveries in lengthy letters to Manson, who became greatly excited. He urged Ross to 'follow the flagellum' in its path through the mosquito, though he continued to think that the parasites were somehow released into the environment rather than transferred directly by the mosquitoes themselves.

Ross was able to see the developing cysts very clearly in a few of his mosquitoes. What he did not see – and in any case would probably not have understood – was what was actually happening among the cells of the parasites. Two types of parasite cells, male and female, were being produced in the blood of his patients. Once they entered the gut of the mosquito, the male cells produced sperm – the mysterious flagella that Laveran had been the first to see. These sperm found and mated with larger female cells. The result was the formation of zygotes, which then developed into the thick-walled structures that he had spotted under the microscope.

Just at the moment he discovered the cysts, he was snatched away from his work and transferred to Kherwara, where there was no malaria, and then to Calcutta. He could find no volunteers in that disease-ridden city because of rumours that had spread through the population about the dangers of English doctors and their experiments. Unable to continue his work on *Anopheles* and humans, he turned in frustration to the disease that had first been found by the Italians, the malaria of birds. Years before, Manson had suggested that he do this, but Ross had ignored his suggestion in his eagerness to pursue the human-mosquito connection.

The malaria of birds was far easier to work with. Because he could feed hundreds of mosquitoes on malaria-infected sparrows, Ross was now able to work day and night. He soon found cysts in some of the mosquito species that he fed on the birds – these mosquitoes were far commoner than the *Anopheles* that transmitted malaria to humans. As they matured, the cysts became full of little transparent rods that were released when he crushed them. He found rods elsewhere in the mosquito as well. They had apparently been expelled from cysts when they burst spontaneously. Like a cautious bloodhound on the scent he discovered that there tended to be more of the rods up near the mosquitoes' heads. But where were the rods going once they left the cysts?

Finally, and entirely by luck, he stumbled on the structures towards which the rods were all homing. These were coiled, transparent organs that occasionally popped out when he pulled the head off a mosquito. Because of his ignorance of mosquito anatomy, he did not at first recognize them. Careful dissection convinced him at last that they were salivary glands, leading directly to the mosquito's proboscis. And in infected mosquitoes the glands were packed with millions of the tiny transparent rods. He realized that the rods, utterly different in their appearance from any stage of the parasite that he or anybody else had seen before, were now poised to enter a new bird host. They would be injected along with the saliva that the mosquito used to keep its host's blood from clotting as it drank. Once he came to this realization Ross was soon able to show that it was these mosquitoes, and not those with rod-free salivary glands, that were able to transmit malaria to uninfected birds.

The year was 1898, and Ross had finally succeeded in closing the mosquito part of the parasite's dauntingly complex life cycle. His success had come in birds rather than humans, of course, but he felt confident that exactly the same thing must be going on with human malarias. As it turned out, he had made his discovery in the nick of time. In the summer of that year the Italians, armed with rumours about his work, began a concentrated effort to find the equivalent human parasites.

Ross, meanwhile, had been sidetracked by the higher authorities in India, who ordered him to investigate kala-azar, a disease that he knew and cared nothing about. (It is caused by a different protozoan, transmitted by the bite of sand-flies, but Ross's first cursory assessment was that it was simply another form of malaria.) He was aching to return to the human malaria that he now thought he understood, and complete the circle of proof using the *Anopheles* mosquitoes that preyed on humans, but he was prevented from doing so.

It did not take the Italians long to home in, as Ross had done, on the *Anopheles* mosquitoes. They showed that the transformations of the human parasite in these mosquitoes were very similar to those that had been found by Ross in *Anopheles* and later in the mosquitoes carrying bird malaria. It was now straightforward to show that *Anopheles* could indeed transmit malaria from one human host to another. But Grassi and his co-workers were frantically rushed, both in the experiments and in the papers that they hastily wrote, because of a fear that they would be scooped by Robert Koch, who had belatedly shown an interest in the problem. Perhaps as a result, their first publication only included a passing reference to Ross's work.

It was this slight that Ross would never forgive – in his memoirs, he never lost the opportunity to refer to Grassi and his colleagues as the 'Italian pirates'. With the same efficiency and single-mindedness that he had brought to the pursuit of the malaria parasite, he tried to line up every luminary in the scientific world against them, and this ploy worked very well. Robert Koch was so swayed by Ross's campaign, and so incensed by Grassi's endless argumentativeness, that he became the prime mover in ensuring that Ross and Ross alone was awarded the Nobel Prize in medicine for 1902.

There has been endless debate since about whether Grassi and his colleagues were unfairly treated. Their work had indeed been proceeding with painful slowness until they were suddenly galvanized by Ross's discovery. And they were less than sensitive to his priority. But they certainly had the resources and the experience to capitalize on his findings swiftly once he had pointed the way. Such a pattern is common in science, and has become even commoner in today's overcrowded scientific world – a postdoctoral fellow or an assistant professor makes a remarkable discovery and is quickly flattened by the steamrollers of large well-organized labs that have been idling their engines waiting for just such a breakthrough. Often the original discoverer does not get much recognition. Ross, at least, fought back vigorously and triumphed. So thorough was his campaign, indeed, that the many important contributions of Grassi, Amico Bignami, Giuseppe Bastianelli and the other Italian scientists, including their seminal finding that malaria can be found in birds, have been largely lost to history.

Fighting the vectors and the parasites

Priority disputes aside, *Anopheles* mosquitoes had finally been fingered as the undoubted villains in the spread of malaria, and of course this opened up all kinds of possibilities for control. Ross thought, on the basis of some crude mathematical calculations, that the draining of marshes and removal of obvious sources of standing water in population centres should be enough to break the cycle. He pushed this idea vigorously during visits in 1901 and 1902 to Liberia and other West African countries. This region was, with good reason, known as the White Man's Grave (and of course it was the Black Man's Grave as well). So prevalent were malaria and numerous other tropical diseases that the exhausted, parasite-ridden inhabitants of the filthy towns Ross visited could hardly be roused to the kind of effort needed to do something about them. None the less, even as Ross argued furiously with recalcitrant local governors, he managed to make some progress. He organized the hauling away of 2500 cartloads of rubbish from the streets and houses of Freetown, the capital of Sierra Leone, and he directed that thousands of puddles on the streets be filled in. Ross recounts (though without hard numbers to back him up) that the health of the European inhabitants improved markedly as a result.

The big early success in malaria eradication, however, came in the New World, as an accidental by-product of an American attempt to stamp out another mosquito-borne disease, yellow fever.

In 1901, US Army medical Major William Gorgas mobilized the well-organized battalions at his command to rid Havana of its immense mosquito population. He employed draconian measures repeatedly draining or oiling every bit of standing water in the city. By October of that year, both the mosquitoes and yellow fever had essentially been

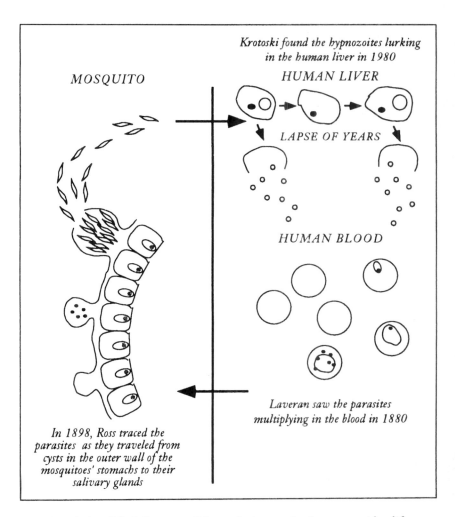

Krotoski found the hypnozoites lurking in the human liver in 1980

MOSQUITO

HUMAN LIVER

LAPSE OF YEARS

HUMAN BLOOD

Laveran saw the parasites multiplying in the blood in 1880

In 1898, Ross traced the parasites as they traveled from cysts in the outer wall of the mosquitoes' stomachs to their salivary glands

FIG. 8-3 A simplified diagram of the malaria parasites' very complex life cycle. The latest link was not found until a century after Laveran's first observation.

driven out. There have been only a few minor recurrences of the disease in Havana in the decades that have followed.

The incidence of malaria was also dramatically reduced. During the period 1898–1900, malaria had actually caused six times as many deaths in Havana as yellow fever. Gorgas's 1901 campaign reduced malaria deaths by ninety per cent. In succeeding years the malaria toll dropped even further. Gorgas was able to achieve a similar success in Panama three years later – yellow fever was swiftly conquered, while malaria dropped to low levels.*

Yellow fever is a far more high-profile foe than malaria, killing its victims in a spectacularly gruesome fashion. This difference may explain the widespread indolence that was exhibited by colonial administrators in countries where yellow fever happened not to be prevalent, particularly when they were confronted by all the hard work necessary to make progress against malaria. Having a mild case of malaria could be something of a badge of honour for colonial administrators, and its effects, it was supposed, could always be warded off by a few gin and tonics at the Officers' Club. The devastating course of the disease in local populations, marked chiefly by high infant mortality and maternal deaths during childbirth, was often ignored.

None the less, the twentieth century has seen remarkable advances in malaria prevention, primarily in regions where the disease is at the periphery of its range. Some were the result of massive alterations of entire ecosystems, like the draining of the Pontine marshes of central Italy. Mosquito larvicides were also introduced, with great success. The first really effective one was Paris green, a poisonous powder made up of a mixture of copper acetoacetate and copper arsenite. It killed mosquito larvae on contact, and in the 1920s and 1930s it was spread liberally on standing water in many tropical and subtropical areas. It probably wreaked a great deal of ecological havoc at the same time, but this was long before the effects of chemical pesticides on ecosystems were understood or could be measured.

After World War Two, the less blatantly poisonous DDT took the place of Paris green. Then DDT was largely abandoned in turn, when it was discovered that it became concentrated in the fatty tissue of birds and animals at the top of the food chain and was destroying the reproductive capacity of some of them.

Direct treatment of malaria victims improved dramatically with the

* Malaria has since rebounded, but cases of yellow fever have not yet resurfaced in appreciable numbers in Latin America – though there has been a slow but steady rise in outbreaks in Africa. The extremely effective vaccine, introduced half a century ago, gives essentially lifetime protection but public health officials fear it is not reaching enough people. And the carrier mosquito populations are increasing. Dengue fever, related to yellow fever, is on the rise in many parts of the tropics, and the fear is that yellow fever may not be far behind. Yellow fever may be another epidemiological time bomb.

development of drugs such as chloroquine and primaquine that affected the parasites' metabolism. Chloroquine and its many derivatives are modifications of a compound called quinoline, found in the bark of the cinchona tree from Peru. It is the active ingredient of the quinine that had, centuries earlier, been found to reduce the severity of malaria attacks.

Quinoline has been in widespread use in one form or another since the sixteenth century. Recently, Andrew Slater and his co-workers at the Picower Institute for Medical Research in New York have obtained a hint of how this compound and its derivatives actually work. Their discovery closes a fascinating loop in time, taking us back to 1880 and Laveran's initial observation of the parasites in soldiers' blood in Algeria. You will remember that Laveran was able to see the almost invisible parasites primarily because of the clusters of dark crystalline granules within their cells.

Like many other cells, these single-celled parasites digest their meals within specialized membrane-lined cavities called food vacuoles. The crystals that Laveran observed were the remnants of those meals, made up of the toxic and indigestible remnants of the haemoglobin molecule. These consist primarily of an iron-rich part of the molecule known as heme. It is the heme that actually carries the oxygen when the haemoglobin is functioning normally in our red blood cells.

The iron in the heme is toxic to the parasites, which is why they sequester it in the form of insoluble crystals. Slater found that the parasites have an enzyme that is responsible for building up these crystals of toxic heme. This concentrates the iron in their food vacuoles and prevents it from poisoning them. The antimalarial drugs, which also tend to be concentrated in the food vacuoles, inhibit this enzyme. When the drugs are present, the iron is free to diffuse through the rest of the parasite cell, killing it.

The effectiveness of these drugs has been much reduced in recent years, however, now that strains of malaria parasite that are resistant to them have appeared. Resistance is popping up more and more rapidly as each new variant of quinoline is introduced – strains resistant to one of the most powerful variants, mefloquine, surfaced in Thailand and in some parts of Africa almost immediately after its introduction in 1985. On top of this, the further the drugs stray from the relatively benign structure of quinoline, the more dangerous they become. I always feel very nervous about taking mefloquine during visits to malarious regions, since it can occasionally cause violent neurcpsychiatric symptoms.

Slater also discovered the mechanism of this resistance. Unfortunately, his discovery bodes ill for the future of quinine-like drugs in the control of malaria. He found that the resistant parasites have somehow acquired the ability to avoid concentrating the drugs in their food vacuoles in the first place. This means that new generations of such drugs must somehow be designed to overcome the ability of the cell to exclude them, while at the same time retaining their effect on the enzyme. It seems likely that succeeding generations of modified quinoline-based drugs will prove to be harder and harder to design without unwanted side-effects.

New and very different drugs are on the way. Artemsinin, derived by Chinese researchers from a fernlike plant that has been employed as an antimalarial in China for at least 1500 years, is now being used widely in Southeast Asia even though neither its effectiveness nor its mode of action have been properly determined. And taxol, the drug obtained from yew trees of the Pacific Northwest that has been used to treat a variety of cancers, is also proving very effective against drug-resistant malaria strains. Taxol interferes with the dynamics of subcellular structures essential for cell division, so that it severely affects the rapidly-dividing cells of the malaria parasite. Unfortunately, it currently costs about six million dollars a kilogram. And there is no doubt that malaria strains resistant to these drugs will arise as well.

Malaria at the ends of the world

The world of our ancestors, whether they lived in the tropics or the temperate zones, was bounded by swamps and marshes. Rivers, untamed by agriculture, regularly overflowed their banks, so that there were networks of meandering streams everywhere. The home of some of my own ancestors, in the eastern part of England, was made up of an almost continuous marsh, different parts of which have been given many names – the Fens, the Norfolk Broads, and the great estuary of the Wash.

In a famous incident, King John lost England's Crown Jewels as he crossed the quicksands of the Wash in the year 1216, and earned centuries of opprobrium. At the time of King John, settlers had penetrated only a little way into these trackless regions, and farmers had to survive on small plots of soil that they had painfully reclaimed from the marshes. These farmers were resigned to the alternating fevers and chills of malaria that almost invariably afflicted them during the summer and autumn.

England? Malaria? Surely not! Any possible connection between the two goes against all that we think we know about the disease. Malaria is an affliction of the tropics and of the subtropics. It is surprising for us to learn that in the Middle Ages, and even in the centuries following, malaria was widespread in England and was found even further north, penetrating into Scotland and Scandinavia. When John Graunt examined the mortality of London, the most frequent cause of death was 'agues and fevers'. Precise descriptions by Graunt's contemporary, the noted doctor Thomas Sydenham, leave no doubt that malaria made up an important fraction of these deaths.

It is difficult to imagine a more unlikely place than northern Europe for such a tropical disease. Cold, brief, rainy summers gave little opportunity for the survival of an organism that must be spread by insects during the warmer months. Whole years must have gone by during which the mosquito vectors were few and far between. How did the malaria parasites survive during all that time?

During the first half of the nineteenth century, malaria was a devastating disease among the poverty-stricken farming communities of Denmark. It was especially prevalent on the islands of Sjaelland, Lolland and Falster, lying to the east of the main Danish peninsula in the channel called the Kattegat that connects the North Sea with the Baltic. A particularly severe epidemic was recorded in 1831, so frightening that it was feared the islands would be depopulated. Almost every inhabitant was affected, sometimes so violently that workers would collapse suddenly in the fields. In one county of the island of Lolland, half the population was affected and many died.

The last such epidemic took place in 1862. Then, dramatically, the disease declined. The medical officer for the area reported only ten cases from the period 1887 to 1900, and after 1900 not a single one.

Today, of course, the islands are intensely farmed. Narrow strips of forest, sometimes only a few feet wide, are all that is left of the woods that once covered them. But a century and a half ago farming methods were primitive, large areas of the islands were still wild, and the farmers' pigs, cattle and horses were free to roam around the countryside much of the time.

These primitive agricultural methods began to change in the second half of the nineteenth century. Domesticated animals were confined more closely in newly-built barns and sties, particularly during the night. And it was just at this time that the incidence of malaria plunged.

When the zoologist Carl Wesenberg-Lund began a survey of Danish mosquitoes early in this century, he was surprised to find that *Anopheles maculipennis*, the species that he knew to be responsible for malaria transmission, was remarkably rare. But then one day in 1918 he happened to enter a cowhouse near his laboratory outside Copenhagen. There he found enormous numbers of *A. maculipennis* hanging in clusters from cobwebs and rafters. Everywhere he went in the countryside, the same pattern was repeated. It seemed that malaria had declined because, as soon as domestic animals were put in shelters during the night, the mosquitoes were able to shift their feeding habits to them and away from humans. And as the mosquitoes changed their feeding habits, they lost their parasites. This was because the parasites could not survive in the farm animals that were now providing most of the mosquitoes' blood meals. The malaria cycle had been broken. These newly parasite-free *Anopheles* mosquitoes, while still plentiful, were no longer vectors of the disease.

This remarkable Danish story, of a public health problem that disappeared before anyone even knew that it was a public health problem, was repeated with variations elsewhere in northern Europe. Although malaria continued to rage along Europe's southern rim, in the north it was on the decline throughout the latter part of the nineteenth century. It seems that the tenuous grip of malaria on the human and mosquito populations of the northern fringes of its distribution could be broken by something as simple as a change in farming methods – even though the change had taken place without any thought of the role it might play in disease control.

This still leaves unanswered the question of how the parasites managed to persist during those long cold northern winters before ecological changes stopped their spread. The answer to that question, as you might guess, is provided in part by the case history of my uncle, who came down with malaria so many decades after leaving the tropics. Some species of the parasites can survive for long periods, not in mosquitoes but in their human hosts.

Up until the middle of the nineteenth century, the arrival of spring in northern and eastern Europe was usually accompanied by an outbreak of malaria among farmers. The first cases appeared even before any mosquitoes had yet emerged from their pupal cases after their long winter's dormancy. By the time the mosquitoes began to appear, the blood of their human hosts was already swarming with malarial parasites, ready to be picked up and transferred by these flying vectors during the brief spring and summer. In the autumn the parasites retreated to their human hosts again. There they eked out another winter's precarious existence in the bodies of the farmers and their families, who were themselves barely surviving among the chilly bogs and fens.

This phenomenon of malarial relapse had been noted by the ancient Greeks and Romans. When, in the first part of this century, a few pioneering researchers turned their attention to the mechanism of this relapse, they soon realized that in order to understand it they had to concentrate on the part of the malaria life cycle that took place in humans. Battista Grassi and his co-workers in Italy had observed odd patterns of relapse among their patients, sometimes after an asymptomatic period that could last for years. In their numerous papers they repeatedly pointed out that these relapses were a strong indication that the life cycle of the parasite in its human host must involve more stages than had yet been discovered, though they did not manage to find proof. But, unfortunately, their shrewd assessment of the available evidence was immediately dismissed and ignored, because of an incorrect observation that was published by a much more famous scientist.

The parasitologist Fritz Schaudinn, of the University of Berlin, announced in 1903 that he had done a simple experiment. He had added Ross's rodlike sporozoites, freshly isolated from the salivary glands of an infected mosquito, to human blood on a microscope slide. There, they apparently pushed their way immediately into red blood cells. The parasite, it seemed, spent almost no time outside the red blood cells of the human host. Both the cycle and the case, Schaudinn claimed, were closed.

FIG. 8-4 A cartoon by Thomas Nast showing the effects of malaria in Washington DC. Nast himself died of yellow fever in Ecuador in 1902.

Schaudinn had already made important discoveries about amoebic dysentery, and was soon to become famous as the co-discoverer of the bacterium that causes syphilis. We do not now know why, since he was such a careful scientist, he made such a series of incorrect observations, but he managed to throw malariologists off the scent for decades. Once again, it was work with bird malaria that eventually led them back to the right track.

In the 1930s, Russian, Italian and German scientists began to find new and unexpected forms of malaria parasites in birds, of a type never seen before. They were discovered in many different tissues. Sometimes they were found in that most dangerous place for a malaria parasite to be, the brain. These parasitic cells looked very different from those that were seen in the blood – large and complex, they often contained a number of nuclei and appeared to be undergoing rapid cell division. But it was not until 1948 that similar parasites were found in mammals.

At the time of these discoveries, Henry Shortt and Percy Garnham were working at London's famous School of Hygiene and Tropical Medicine, located in the middle of the nest of hospitals and university buildings that surround Gower Street in Bloomsbury. They examined the liver of a macaque monkey that was undergoing a malarial relapse, and found that some of the liver cells harboured alien cells within them. These parasites looked very much like those that had been seen in birds.

The obvious question was whether similar cells could be found in humans as well. If humans responded to malaria like monkeys, the most likely place to look would be in the liver of a patient afflicted with a massive dose of malaria. Luckily, Shortt and Garnham did not have to travel to the tropics to find such a patient – one was literally about to be created for them, in the form of an inmate at a mental hospital in Aylesbury, near Oxford. The inmate suffered from advanced syphilis.

Early in this century, Julius von Wagner-Jauregg, a neurologist at the University of Vienna, had made a remarkable series of observations. He noted that patients with paralytic syphilis who also happened to come down with malaria infections sometimes showed a dramatic improvement afterwards, with a lessening of paralysis and increased mental abilities. Apparently this was a result of the high fevers that the malaria induced. Wagner-Jauregg then took the daring step of deliberately giving his patients malaria, and achieved some astounding recoveries. The attack of malaria that he had induced could of course be cured in its turn by quinine – in those distant days the malaria parasites had not yet become sophisticated enough to resist this centuries-old prophylactic.

Malaria therapy was soon in widespread use, and helped thousands of syphilitic patients. For inventing this unique method of using one disease to fight another, Wagner-Jauregg received the Nobel Prize in 1927.

Remarkably, even after the advent of penicillin, Wagner-Jauregg's treatment was still being used in England in the late 1940s. While Shortt and Garnham stood by, a container full of hundreds of mosquitoes carrying *P. vivax*, the relapsing malaria, was pressed to the skin of the Aylesbury patient. For good measure, other mosquitoes had been dissected to isolate millions of sporozoites, and the resulting brew was injected into the patient's veins. A week later, shortly before the malaria symptoms began to appear, the patient volunteered to undergo a liver biopsy. As Shortt and Garnham had expected, his liver was found to be filled with the mysterious alien cells of the parasites, presumably poised to multiply and enter the bloodstream.

So the malaria parasites could indeed lurk unsuspected in the body, as Grassi had supposed half a century earlier. It now became apparent that Ross's sporozoites, entering the bloodstream from the bite of an infected mosquito, do not immediately attack the red blood cells as Schaudinn had claimed but rather are carried by the circulation to the liver. There they burrow into liver cells and, safely sequestered, grow into the large intracellular parasites that Shortt and Garnham had found.

But residual questions remained. Why should some of the cells lurking in the liver produce offspring immediately and others wait for months or years or even decades before taking the plunge? Were there two – or perhaps more – types of these cells in the liver, some scheduled to 'go off' early and others waiting like ticking time bombs for who knew what kind of signal? Or did some of the cells divide, enter the bloodstream, then think better of it and return to their protected life in the liver? Shortt and Garnham thought that such a hesitant career for some of the parasites was likely, because they had found that some of the parasite cells in the liver of their volunteer were still dividing after three and a half months. But this was weak evidence at best, and later Garnham became convinced that this supposed indecision on the part of the parasites was not the right answer

It was not until 1980, a full century after Laveran's original observations, that a chance observation led to the finding of another piece of the puzzle. Wojciech Krotoski, a medical scientist working in the US Public Health Service hospital in New Orleans, had developed new ways to visualize the malaria parasites by labelling them with fluorescent antibodies. He found that he could use this technique to follow the early stages of the invasion by Ross's sporozoites, which were otherwise very hard to locate in the mass of material from a typical biopsy. The labelled parasites were illuminated brilliantly when he shone ultraviolet light through a preparation of liver cells from a patient.

He happened to show a slide containing one of these large cells, fluorescing brightly, to another member of the laboratory. His co-worker's attention was immediately distracted away from the large cell that Krotoski was proudly showing off. 'What's that small bright object at 4 o'clock?' he asked. Krotoski thought at first that it must be some kind of artifact, but he quickly saw that there were many of these tiny fluorescent objects, mixed in with the large bright cells. They must, he realized, be cells derived from Ross's sporozoites that had buried themselves in the liver but had not grown. They had stayed small, in a kind of stasis, but presumably poised to respond to some kind of a signal that would tell them to start growing.

At first, Krotoski called these little cells *dormozoites*, sleeping little animals, but this mixture of Latin and Greek offended those who worry about such things. The equivalent all-Greek term *hypnozoites* was soon substituted. Hypnozoites have now been found in most of the relapsing malarias. And their incidence is correlated with the difficulty that each strain of these parasites faces in surviving the winter. Strains of *P. vivax* from temperate zones have been found to produce more hypnozoites in the livers of their victims than those from the tropics. If it is indeed hypnozoites that cause relapses, then this is exactly what one would expect from a parasite that must somehow survive those long cold winters when the mosquito host is not available.

So, the relatively quick relapses of the kind that my uncle suffered in India had finally been explained. They are caused by hypnozoites. Could these sleeping cells have been woken up by the plunge of temperature that he experienced during his company's manoeuvres in the Siwalik Hills? Malaria attacks can often be brought on by sudden shocks of various kinds, but the underlying trigger itself remains a mystery to this day.

Of course, this still leaves unsolved the attacks that prostrated him in England nearly half a century later. The only malaria known to bring about such belated relapses is caused by *P. malariae*, but no hypnozoites have been found in the livers of people who are infected with this parasite. Handwaving explanations have been suggested. Perhaps parasites of *P. malariae* manage to lurk unnoticed somewhere in a remote corner of the circulatory system, where they can barricade themselves and resist treatments with antimalarial drugs. But without experimental data these explanations carry little conviction. They must lurk somewhere, but there are hundreds of hiding places in the human body. It seems that, even after more than a century of intensive investigation, the resourceful malaria parasites have not yet yielded up all the secrets of their baroque life cycle.

It is striking and important that these tiny hypnozoites have never been found in the most severe of human malarias, that caused by *P. falciparum*. And this, of course, is the malaria that does not relapse.

Parasite family trees

Much remains to be understood about these fiendishly complex parasites, but now we can begin to build up a picture of how they work, and to probe for weak spots in their defences. We must look at what the various kinds of malaria that cause so much suffering world-wide can do and cannot do. A pattern is emerging. The sophisticated malarias are able to spread beyond the tropics, while *P. falciparum*, the less sophisticated species, is prevented from doing so. And further, just like the plague bacilli, the abilities that it lacks can actually force it to be more severe in its effects.

The malarias caused by *P. vivax* and *P. malariae*, while they can be severe, are not usually life-threatening if care is available. They are known among malariologists as *benign* malarias, though to anyone who has suffered through an attack the name must be viewed as a rather unfunny little joke on the part of the doctors who bestowed it. But the malaria caused by *P. falciparum* can have disastrous effects on its human host, not because it can do things that the other malarias cannot, but rather because it is unable to do things that they can.

Continual reinfection is essential, since the parasites cannot form hypnozoites in the liver. As a result, these parasites cannot use humans as a refuge during periods when there are no mosquitoes. If a victim of falciparum malaria is kept away from mosquitoes, then the parasites will disappear from his or her bloodstream in a few weeks. And this is why falciparum malaria is essentially confined to the tropics and the subtropics, with their year-round mosquito populations.

We do not know whether falciparum malaria has lost the ability to make hypnozoites, or whether it never had that capacity. It will be exciting to search for hypnozoite genes on its chromosomes, to see if they have been inactivated like the invasin gene of *Yersinia pestis*. While this particular search has not been tried, the genes of the malaria parasites have already begun to give up their secrets. We are now beginning to see where the parasites themselves may have come from.

The parasites of the various malaria species have not diverged very far from each other in the course of their evolution, so that their genes are very similar. It is possible to compare some of these genes from one parasite to another. One such gene specifies a part of the parasite's ribosomes, which are tiny structures that manufacture proteins in the cell. Thomas F. McCutchan and his colleagues at the National Institute of Allergy and Infectious Diseases have sequenced and compared this gene isolated from a large number of different malaria parasite species.

The trick in such comparisons is to use the sequence information to build a kind of tree, with the various sequences occupying the tips of the branches. Because some of the parasites are closely related to each other, some of these gene sequences are more closely related than others, and these sequences will occupy branches near each other on the tree. Sequences of more distantly related species may lie at the tips of very different branches.

The base of the tree represents the gene in the common ancestor that has given rise to all the divergent genes in the present-day malarias. Of course, that ancestral species has long since disappeared. The sequence of the ribosomal RNA gene that was possessed by this ancestor cannot be obtained directly, but can only be inferred by comparing the present-day sequences and using them to construct a sequence that is likely to resemble that of the common ancestor.

The method has worked very well for many groups of species of animals, plants and micro-organisms. Species that we know to be closely related, like humans and chimpanzees, have gene sequences that are separated by a far smaller number of differences than species that are less closely related, like humans and the New World spider monkeys.

When McCutchan built trees to find out how the malaria parasites are related to each other, he was able to show that vivax malaria is closely related to the malarias that affect monkeys and apes. At the other extreme, the trees showed that the gene of *P. malariae* has only a remote resemblance to that of any other known malaria. This parasite, it seems, may have had a very long history of infecting our ancestors, so long indeed that it has diverged greatly from all the other malarias. Perhaps this is why it is so clever that it can lurk in our bodies for decades at a time.

The trees also showed that the severe *P. falciparum* parasite has had a very different history from either *P. vivax* or *P. malariae*. It seems that falciparum malaria has spent most of its history preying on creatures quite different from us. Its ribosomal RNA gene resembles those of the malarias that affect birds, which means that it is likely to be a johnnie-come-lately to our species. It is interesting that the malarias of birds do not seem to be able to form hypnozoites, so in this feature as well as in their genes they resemble falciparum malaria.

Another indication of how poorly falciparum malaria has adjusted to its new hosts is that the disease it causes is so much more severe than those caused by the other malarias. Its parasites swarm in the blood-stream in far greater numbers, and these numbers constitute an immediate threat. As they multiply in synchrony, huge numbers of red blood cells are broken open simultaneously. This gives rise to the extremely violent fevers and chills that mark the disease.

The wreckage of the destroyed blood cells can block tiny capillaries, sometimes causing them to burst. It is this process that releases the parasites into the brain, causing the dreaded inflammation of cerebral malaria. In pregnant women who are infected, the detritus from the exploded red blood cells can also damage the delicate tissues of the placenta that nourishes the child in the womb. Again, this produces inflammation and tissue destruction, which can terminate the pregnancy. As a result, malaria-induced abortions are very common in the tropics. In the last stages of a really severe infection, the structure of the kidneys is largely destroyed and haemoglobin passes directly into the urine. This is a sign of the usually fatal blackwater fever.

None of these dangerous symptoms is usually found in the benign malarias (though they can occur occasionally, as they did among the highland Indians of Peru when they were first exposed to vivax malaria). Why should falciparum malaria be so vicious? Is it simply that it has not had time to adapt to us, since we have only recently picked up this form from birds, or are there other reasons? To explore this, we must look at how both humans and mosquitoes protect themselves against malaria.

Losing and winning the war between hosts and parasites

As we saw earlier, about two million people die every year of the direct effects of malaria, and there is no doubt that these parasites contribute indirectly to many other deaths. At least half of the deaths occur among children, who have had little time to develop immunity. Adults who live all their lives in the presence of falciparum malaria and suffer repeated infections (in some parts of West Africa, people may be bitten by hundreds of potentially infected mosquitoes in a single night) do gradually manage to build up a sketchy kind of resistance. The process takes years, because even the *falciparum* parasite spends a good deal of its time hiding inside cells of the liver and bloodstream where the immune system cannot see it. And during the brief periods when it is exposed, it is continually changing its shape and structure. New proteins are always being presented to our immune system. It takes a long time for us, confronted with such a cunning foe, to build up enough antibodies to keep the parasites in check.

Even in those with a high degree of immunity the parasites can continue to multiply, though not in the same overwhelming numbers as in people who are immunologically naïve. If nothing else, at least the immune system can slow the build-up of those pullulating hordes of parasites, so that they are less likely to cause the ghastly cerebral, kidney and placental complications.

It is this difficulty of raising antibodies that has made it so hard to produce a malaria vaccine. Manuel Patarroyo, a Colombian biochemist, has recently manufactured a promising vaccine by combining bits of genes from three different stages of the malarial parasite. The result is a 'chimaeric' gene that makes a hybrid protein. The protein is harmless to humans, but quite effective at stimulating our immune systems to produce anti-malaria antibodies.

Although the first clinical trials of these antibodies were flawed, more recent trials show clearly that his vaccine can reduce the incidence of malaria infections by about fifty per cent. This is a real triumph, for

Patarroyo's vaccine is the first that has exhibited any effectiveness against one of the many parasitic diseases that are caused by protozoa. It is hardly surprising, in view of the complexity of the parasites, that it is not fully effective. And it remains to be seen whether, as many fear, the parasites will quickly evolve new proteins that can overcome the vaccine.

This is because, unfortunately, the parasites do not stand still while we accumulate antibodies against them. New strains of parasite that are less visible to our immune systems are arising continually by mutation. They are more likely to do so in a big long-lived animal like ourselves than in a small ephemeral creature like a mouse, since in our bloodstreams the population of parasites can be huge and the possibility that a protective mutation will arise somewhere among them is correspondingly great.

There have been many remarkably ingenious attempts to circumvent this problem. In the 1950s, the parasitologist Clay Huff of the Naval Medical Research Institute in Bethesda made an unusual observation in the course of investigating bird malaria. He isolated malaria parasites from chicken blood, concentrated them, and used them to raise strong antibodies in birds that had not previously been exposed to malaria. The idea, of course, was to see whether they would resist subsequent malarial infection. Unfortunately, when he gave malaria to these immunized birds, he found to his disappointment that they came down with the disease almost as readily, and produced almost as many parasites, as birds that had not been immunized.

Yet when he tried to use mosquitoes to transfer these cases of malaria from the immunized chickens to uninfected chickens, he had little success. Their blood swarmed with parasites, and the mosquitoes certainly picked them up, but something happened to the parasites during the transfer from one chicken to the next. For some reason once they entered the stomachs of the mosquitoes, the parasites were unable to go through the complex series of transformations that resulted in sexual reproduction and the subsequent swarms of rodlike sporozoites.

Later work, chiefly by Richard Carter of the University of Edinburgh, established that the parasites were actually being destroyed in the mosquitoes. Carter found that it was the antibodies in the blood of the animal host, picked up by the mosquitoes in the course of their blood meals, that were doing the damage. Although the antibodies were not powerful enough to destroy the parasite cells when they were in the blood of the chickens, they were able to cause enough damage to them when they had been picked up by the mosquitoes to prevent them from being transformed into sex cells so that they could complete their life cycle.

Carter then suggested an utterly novel way to prevent the spread of malaria. The Sisyphean task of immunizing humans against those ever-changing parasites was proving to be frustratingly difficult. But suppose one tried to cure, not the humans who had the disease, but the mosquitoes! Unlikely as it sounds this actually might be possible, because of the fact that the parasites go through such dramatic changes in their passage through the mosquitoes. New proteins are synthesized on the surface of their cells, proteins that are never made by the parasites when they are in their human hosts.

Suppose, Carter thought, one were to deliberately inject people who already have malaria with a parasite protein that is expressed only when the parasites are living in mosquitoes. No good (or presumably harm) would be done to the people who receive the shots. But once the parasites enter the mosquitoes and begin to manufacture that protein, then they might be irreparably damaged by the antibodies that accompany them in the host blood. The mosquitoes could drink as much of the blood of these infected people as they liked, but they would no longer be able to transfer the disease to people who did not have it. The life cycle of the parasite would be broken, not in the humans, but in the mosquitoes.

This unlikely-sounding idea, which Carter called *transmission-blocking immunity*, is now actually being vigorously pursued. Two groups of malaria workers based in London, one headed by Geoffrey Targett of the School of Hygiene and Tropical Medicine and the other by Robert Sinden of Imperial College, are trying to find the ideal antibody-producing protein. Targett is working on proteins found on the surface of the gametes, and Sinden is concentrating on those that are found on the surface of the zygote that forms in the mosquito's stomach wall after the gametes have fused.

Will transmission-blocking immunity work? Sinden is about to start a pilot project to try introducing his best zygote protein into East African volunteers. He emphasized to me that, like all other attempts at malaria control, transmission-blocking immunity will have to be carried out on a massive scale if it is to do any good. The parasites, of course, will fight back – their existence depends on it. It is not likely to be long before mutant parasites arise that carry altered proteins on the surface of their zygotes, proteins that will be invisible to the antibodies that have been raised to their older versions. Of course, these new mutant proteins can also be isolated and purified, a task made easier by new recombinant DNA techniques, but the process is still likely to be time-consuming and expensive.

Other high-tech approaches, although they are currently at or beyond the limits of what might be possible, hold promise for the future. One of the most exciting discoveries in genetics in recent years has been the finding of little transposable elements called P elements in the fruit fly *Drosophila*. Even though they actually cause harm to the flies that carry them, these elements can spread readily through a population of flies. This is because they make many copies of themselves, which are inserted into many places in the chromosomes of their hosts. When a P element invades a fly, its chromosomes can break out in a rash of these elements, a kind of chromosomal disease. Thus, even if a fly receives only one copy of a P element on one of the chromosomes it receives from its parents, the element can multiply and infect all the other chromosomes. This ensures that copies of the element will be passed on to all the fly's progeny. The elements can even hop from flies of one species to those of another, by mechanisms that are not understood.

This opens up an exciting possibility. Suppose that such elements could be designed to spread through a population of *Anopheles* mosquitoes, and at the same time to carry genes for resistance to the malaria parasite. Then, if mosquitoes carrying these elements were to be released in large numbers, this might cause the element to spread through the whole immense population of mosquitoes in Africa, wiping out the parasite in the process.

The construction of such elements is a long way from realization, however, and such expensive and technologically difficult programmes are unlikely to be carried out successfully or to completion in the Third World. In the meantime the slaughter caused by the *falciparum* parasite continues.

Attempts to eradicate malaria by less sophisticated means have generally proved to be frustratingly difficult. The population of Sri Lanka has always suffered from heavy infections of both vivax and falciparum malaria. This has been exacerbated by the introduction of intensive agriculture, which has allowed the mosquito vectors to breed in immense numbers. The worst recorded year before World War Two was 1935. In that year, malaria swept the island – in the interior town of Anuradhapura deaths reached a peak of 64 per thousand inhabitants. The toll was probably even higher, since many cases of malaria were misdiagnosed as other fevers. Then, in 1945, the Malaria Control Programme was begun under the auspices of the World Health Organization. All the stops were pulled out – DDT was sprayed heavily, there were house-to-house searches for places where mosquitoes might breed, antimalarial drugs became widely available, and a greatly improved programme of detection and hospital treatment was introduced. Just as in Peru, falciparum malaria was the first to disappear, and soon the new cases of vivax malaria plummeted. By 1955, there had been a 100-fold reduction in the number of cases. And astonishingly, by 1964 there were just 17 new cases of vivax malaria, and none of falciparum, on the whole island.

The programme began to run into difficulties in the 1950s as Sri Lanka's economy started to suffer from the effects of overpopulation and the ruling government was distracted by rising nationalist movements. Spraying and eradication efforts were largely abandoned by the mid-1960s, and malaria rebounded almost instantly to half a million cases in 1969. It is telling, and suggests something about the relative vigours of temperate and tropical malarias, that the temperate-adapted *P. vivax* was the first to reappear, followed a decade later by the tropical *P. falciparum* – even though Sri Lanka is of course a tropical island.

Malaria has remained ever since at levels that sometimes approach those of the early 1950s. Luckily, some vestiges of the public health programme remain, and while the incidence of morbidity – severe and damaging cases of malaria – has shot up, actual mortality directly caused by the disease has remained relatively low.

What is remarkable about this otherwise depressing story is that malaria eradication was so nearly successful. A sturdier political will, coupled with the determination to see the job through to the end, might have actually produced the first real conquest of malaria in a tropical region.

A similar trajectory was followed in India. In 1964, at the height of the eradication campaign, there were only about 10,000 new cases of malaria in the whole vast country, and no deaths. But then, in 1976, a 'modified plan' of operations was begun, and malaria cases began to rise once more. When I visited Vellore, I was told that the town had been considered for years one of the great success stories of the Indian malaria eradication programme. It was officially declared malaria-free. Only recently has the government been forced to admit, in the face of overwhelming evidence, that malaria, including falciparum malaria, has reappeared in Vellore.

The conquest of malaria was almost a trivial matter in the far north of Europe, where a slight change in the living habits and practices of farmers was enough to break the cycle. It is easy to see why the cycle has been so much harder to break in the tropics, for a large number of factors conspire against malaria eradication. Mosquito populations are large and breed year-round, and often several different species of mosquito are malaria transmitters. The human populations tend to be poor, and to live in houses that afford them no protection. (The incidence of malaria among people living in houses without walls in Sri Lanka is twice that of people who can afford more substantial housing.) And falciparum malaria produces immense numbers of parasites that mosquitoes can easily transmit to new hosts. Yet, as the tantalizing experience of Sri Lanka shows, it remains possible that malaria might even be conquered in the tropics.

One reason for this potential vulnerability is that the parasites do not have it all their own way even in these warm, moist regions. Why does the *falciparum* parasite, in particular, have to multiply in such huge numbers in its hosts' bloodstreams, putting them into such danger? The usual explanation for the existence of overly dangerous parasites is that they are new arrivals in our species. Such parvenus, says this received wisdom, tend to overdo things because they are on unfamiliar terrain and have not had time to adjust to their new host. Yet it is likely that the *falciparum* parasite has coexisted with its human hosts for thousands of years – one would surely think that it would take much less time than that for strains to be appear that are less hard on the humans they infect. The other malarias multiply in the bloodstream to far smaller numbers, but they can still make quite enough gametocytes to complete their life cycles. Why is falciparum malaria trying so hard?

What problem or problems is it trying to overcome? Are there pieces to the puzzle that we have not yet found?

Remember that there is strong evidence that falciparum malaria is a recent arrival in our species from birds. And it may also be a recent arrival in the *Anopheles* mosquitoes that prey on us as well, since most bird malarias are carried by mosquitoes of the genus *Culex*. If this is so, there may be real obstacles to the survival of falciparum malaria in both of its new hosts, human beings and *Anopheles* mosquitoes. These obstacles may have caused it to react in ways reminiscent of the immense overproduction of plague bacilli that we encountered in Chapter Four. Consider just a few of the difficulties facing the parasite.

First, it cannot make hypnozoites in the liver. This option is cut off, just as the option of entering the cells of the gut lining is cut off for the plague bacillus. So the parasites must escape to a new host during a short span of time. A host who has not been reinfected by a new wave of *falciparum* sporozoites from further mosquito bites will lose his or her parasites, recover (with luck), and become completely noninfectious.

Second, the parasites must be able to complete the mosquito part of their life cycle. This may be more difficult than we think. It is striking that in subtropical areas the *falciparum* parasite always seems to be easier to knock down than the other malarias. It was the first of the three prevalent malarias to disappear during the enormously successful eradication programme that was carried out in Taiwan in the late 1950s. Even in the much less successful programme that was started in Peru somewhat later, it was forced down to very low levels. It is only in the last few years, with the emergence of drug-resistant strains, that it is on the rise again. And it only reappeared in substantial numbers in Sri Lanka several years after the reappearance of *P. vivax*.

There are few things that occur to biologists that Mother Nature has not thought of first. It may be that transmission immunity is already operating in the anopheline mosquitoes, without the aid of scientists. This may be at least part of the reason for the difficulty that *falciparum* seems to encounter in spreading – and may also account indirectly for the viciousness of the parasite. Years of field work carried out in The Gambia and elsewhere give some support to this idea.

The Gambia is a swampy and poverty-stricken little country on Africa's west coast, where slavery was only officially abandoned in 1906. The region has one of the highest incidences of malaria in the world.

Anopheles mosquitoes of a dozen different species swarm year round, and all four types of human malaria are present. Thousands of children come down with cerebral malaria each year, and of those that recover at least a quarter suffer permanent brain damage.

The picture that has emerged in The Gambia, as Geoffrey Targett sketched it for me, is one of astounding complexity. There are indeed large differences in the susceptibility of children to the effects of malaria, and in the likelihood that they will pass the disease on, but the extent to which these differences are due to transmission immunity remains unclear. There is genetic variation in the populations of the parasites themselves and in the populations of the mosquitoes that carry them. Some strains of a mosquito species will allow the parasites to develop normally, while others will only allow a few zygotes to mature. To make things even more complicated, some people have been shown to be far more attractive to anopheline mosquitoes than others. Variation in hosts and parasites affects every stage of the cycle Yet Richard Carter has found traces of antibodies to the sexual stages of the *vivax* parasite in patients in Sri Lanka, so the potential for transmission-blocking immunity is a real one.

Whether or not transmission immunity is operating, mosquitoes are reluctant hosts for *P. falciparum*. When *Anopheles* mosquitoes are trapped at random, most are found to be free of the parasites, even though all of them are fully capable of picking them up from humans. And in one recent study, while *Anopheles* could be readily infected with *P. vivax* by feeding them on patients with vivax malaria, far fewer of these mosquitoes picked up parasites when they were fed on patients with falciparum malaria. This was the case even though the bloodstreams of the patients with falciparum malaria swarmed with far larger numbers of parasites.

In spite of this, if there are enough mosquitoes around throughout the year the *falciparum* parasites can still manage to be passed from host to host. In West Africa the greatest concentrations of *Anopheles* are found in farmland. These mosquitoes have adapted well to life in the puddles that abound on farms and in the uncovered water containers that are common in the farmhouses. In the cities surrounded by this farmland, however, the mosquitoes are much rarer – and so is the incidence of disease transmission. In one study carried out in the city of Kinshasha, Zaire, there were less than 0.1 infective bites per night per person, one-thirtieth the rate in the nearby countryside. One reason for the difference was that only one species of mosquito capable of transmitting malarias lived in the city itself, compared with four in the countryside.

Whether the parasites' difficulties are due to transmission immunity or simply the result of a poor adaptation to their mosquito hosts, it seems that *falciparum* parasites even in their native tropics are living on the edge of what is possible. Can something be done to capitalize on their apparent vulnerability? In particular, how low does the transmission rate need to be driven before the cycle is broken? How much do people's habits need to be changed? Already, in West Africa, something as simple and low-tech as the introduction of pyrethrin-impregnated mosquito netting has had a substantial impact (though, alas, the pyrethrin turns out to be a carcinogen). And slight alterations in the environment can have unexpectedly large effects. Farmers do not need to migrate to cities to gain a health benefit. In The Gambia, childhood mortality from malaria was found to be much lower in larger villages than in smaller ones as the effect of urbanization on mosquito populations began to be felt.

It is ironic that these parasites are so vulnerable to environmental changes that there is a large impact on the cycle of infection when people migrate into sprawling slum-ridden cities like Kinshasha. Of course, one danger may simply have been exchanged for another, since the overcrowding and insanitary conditions of these new urban areas expose people who live in them to outbreaks of diarrhoeal diseases that may be even more dangerous. But in general, as with so many other diseases, it seems likely that the cycle of infection can be broken with relative ease even in the tropics. A combination of destruction of wild habitat, spraying, a change in living habits, immunization (including the employment of transmission immunity), genetic engineering, and the wide availability of antimalarial drugs, taken together, may eventually make it impossible for the parasite populations to maintain themselves. If this happens, I predict that the most dangerous parasite, *falciparum*, will be the first to succumb. It is the one that could not escape from the tropics, and it is the one that is least likely to escape the onslaught of technology.

Like the strains of typhoid that have managed to leave the tropics, the malaria species that have done so are more sophisticated than the one that was left behind. The long history of coexistence of both typhoid and malaria with humans has led to the evolution of ingenious mechanisms for their survival in such unfavourable environments. Typhoid in the temperate zones is able to hide itself very effectively from its victims' immune systems, and to exist in unexpected interstices in the body for long periods. Temperate-zone malarias can conceal themselves in the liver for long periods, and perhaps elsewhere as well – recall the ability of *P. malariae* to survive for decades. And they seem far more adept at surviving in their mosquito hosts. The plague bacillus, on the other hand, is a new arrival in our species, and perhaps in a number of other mammalian species as well. Its adaptation to transmission by fleas and its requirements for other human hosts to be nearby have led it to be multiply-crippled compared with its free-living relatives.

The degree of sophistication of a pathogen's adaptation is correlated in part with how long it has survived in our species and in part with its mode of transmission. Notice that the malaria parasites do not need a crowded human population to be transmitted from one person to another – the mosquitoes are quite capable of carrying them for fairly long distances. This means that malaria can be endemic in the temperate zones for long periods, which will give ample opportunity for sophisticated evolutionary mechanisms to arise. The plague bacillus, in contrast, depends for its survival on the crowding of its hosts. Such crowding is less common, the window of opportunity is smaller, and the evolutionary mechanisms that have arisen are correspondingly cruder.

Sometimes, pathogens are hemmed in by even more restrictions. They may, to survive, have to search out a bit of the tropics in the temperate zones.

9

Syphilis and the Faustian bargain

The time has come to turn the light on an existing
condition touching the American fireside that will
appall the average woman and girl.

Edward Bok, editorial, *The Ladies' Home Journal*, 1908

When Christopher Columbus returned to the island of Hispaniola on
his third voyage on 3 August 1498, he found everything in confusion.
He had left his brother Bartolomé there as viceroy in March 1496, and
during the two and a half years of his absence Bartolomé had cruelly
oppressed the Indians. There had also been a rebellion among the
Spaniards themselves, led by Bartolomé's lieutenant Francisco Roldán,
though Roldán quickly made peace when Columbus's advance vessels
arrived. And to add to the gloom and depression of the occupying
Spaniards, every third member of Bartolomé's little band was suffering
from syphilis and an assortment of other diseases.

Hispaniola was anything but hospitable to the invading Europeans.
On his very first voyage in 1492, Columbus had left some men behind
on the island, in an attempt to establish a settlement. During his second
voyage a year later he found this nascent colony gone, wiped out by
the natives of the islands. There is no record of disease having played
a role in the failure of this first attempt, though it may have done. When
Columbus departed for Spain for the second time, still determined to
found a permanent colony, he left behind a far larger number of men.
They were better equipped to fight off the natives, though not the
diseases. There is no doubt that illness, dissension and mismanagement,
coupled with the growing rage of the native population, would have

destroyed this second foothold of the Spaniards in the New World as well, had Columbus's return been delayed by even a few months.

Columbus managed to smooth things over with Roldán and reassert some limited authority. But his little empire was still in sad disarray when Francisco de Bobadilla, an emissary from the court of Ferdinand and Isabella, arrived two years later. Bobadilla had been charged by the Spanish monarchs, who were receiving distressing reports of Columbus's ineptitude, with finding out just how bad things were. Numerous conflicts between Bobadilla and Columbus eventually led to Columbus's arrest, and to his being taken back to Spain in chains and in disgrace. He was soon pardoned, though not restored to his former powers, and such was his reputation that he was even able to gather resources to finance a fourth voyage.

Long before this time Columbus's mental health, perhaps never robust, had begun to fail. Signs of this had appeared at the outset of his third voyage. Before his arrival in Hispaniola, he had explored the seas around the island of Trinidad and found signs of fresh water flowing from the mouth of the mighty Orinoco River. He promptly supposed that he had discovered one of the four rivers of Paradise, which flowed from the Garden of Eden. To explain the Garden of Eden's location in that rather unlikely place, he made the further assumption that the world was actually pear-shaped rather than round. The Orinoco was at the top of the pear, and was therefore the nearest part of the planet to heaven.

Was Columbus himself suffering from syphilis, and does this explain his progressive mental derangement? Certainly, when he came back to Spain from his final voyage at the end of 1504, he was clearly mentally ill and his legs were paralysed – though of course there are many other possible causes of these symptoms. At this time he was only 53 years old. He was still alert enough to dictate his will on 19 May 1506, two days before he died.

Columbus had brought ten of the inhabitants of Hispaniola back with him on his first voyage, and in 1495 he officially initiated the slave trade between the Old and New Worlds by sending five hundred natives back to Spain in a ship commanded by Antonio de Torres. About three hundred of them, whom Columbus had called rebels rather than slaves in order to assuage his conscience, survived the trip. And, though Columbus had been the only one brave (or perhaps crazy) enough to lead the way, by the time of his death the waters of the Atlantic were filled with other ships packed with adventurers eager to exploit the new world he had found. The dreadful traffic in human cargo, and the inevitable transfer from the Old World to the New of smallpox, measles, mumps and many other diseases, had begun.

But was there a transfer of disease from the New to the Old World as well? Did Columbus introduce syphilis to Spain?

The controversy over this possibility has raged down to the present time, and still remains unresolved. It forms one of the most complex, puzzling and fascinating medical detective stories of all time. I cannot attempt to resolve it here (though I have preferences that will become apparent as the story unfolds). But even to try to understand the story, we must begin by realizing that syphilis is one extreme of a continuum of diseases, most of them surprisingly little understood, which have afflicted humans and their relatives for a very long time. Syphilis is simply the most extreme manifestation of the attempts of a parasite to spread from host to host. Like so many other diseases that have managed to spread beyond the tropics, it has found new and ingenious ways to survive, but at a cost to itself as well as its hosts. Like the plague bacillus, it is the result of a series of evolutionary compromises that have opened up a risky new niche.

Penicillin, first used against infections in 1943, has been extremely effective against syphilis and its allied diseases. Doctors of today are able to cure the infection in its earliest stages, and are not confronted, as were doctors in the 1930s and earlier, with patients in whom it had run riot for decades. Occasionally we get glimpses of how devastating its course used to be. The notorious Tuskegee Study, of men deliberately left untreated for the disease, was revealed to a shocked world in 1972. This cruel and unnecessary study had been carried out over a period of decades by doctors in the US Public Health Service, and yielded nothing of value.

Until the appearance of AIDS there was no disease quite like syphilis. To the world of the early explorers, it represented a direct punishment for sin, a clear sign of how far we had fallen from grace since the banishment of Adam and Eve from the Garden of Eden. The disease has been a source of horror and fascination over the centuries, because of its multitude of repellent manifestations and its mode of transfer. The connection between syphilis and sex, particularly illicit sex, which was clear from the moment of its first appearance in Europe, injected a new element of terror and guilt into one of our most basic drives.

At the present time, each sexual contact with an infectious person yields about a thirty per cent chance that the disease will spread. Syphilis begins relatively mildly, after a two- to six-week incubation, with the formation of a small painless ulcer or chancre on the genitals. It is during this period that the disease is at its most infectious. Then the chancre fades away, and for a brief time the victim imagines that he or she is cured. Six to eight weeks after the appearance of he chancre, however, a new set of symptoms appears, lumped together under the term secondary syphilis. In different people these symptoms can range from the mild to the horrific – fever, headache, and an astoundingly variable collection of spots, lesions and pustules on the surface of the body. These symptoms too slowly disappear over a period of months. In a minority of cases, even without treatment, the body's defences seem to work perfectly, wiping out all obvious traces of the disease although infection still persists. But about two-thirds of the time the disease moves on to the next and most insidious stage.

Tertiary syphilis, often marked by tumours in the skin called gummata, can develop over decades. If untreated, these tumours begin to work changes deeper in the body, eating away at the bone in specific and characteristic ways. It is these skeletal changes that allow a diagnosis of syphilis, or of a closely allied disease called yaws, to be made even in centuries-old skeletons. Unlike many other diseases that affect bone, syphilis does not repress the body's ability to heal itself. If a bone is eaten away by the lesions of syphilis or yaws, new bone is deposited on the side opposite the site of the lesion, so that the bone remains strong but becomes progressively more and more deformed.

Tertiary syphilis affects many parts of the body, and one of its most devastating effects is on the spinal cord and the brain. Paralysis, particularly of the face and lower limbs, often develops, caused by infection spreading to the posterior processes of the spinal cord. This is marked by a loss of co-ordination and a characteristic straddling walk. Blindness can result from infection of the optic nerve. In the final stages lesions begin to appear in the brain leading to insanity, paralysis and death. The list of victims of tertiary syphilis is long, and includes some of the stellar names of history and the arts – Francis I of France, Charles Baudelaire, Friedrich Nietzsche, Heinrich Heine, Guy de Maupassant,

Toulouse Lautrec, Jules Goncourt, Gaetano Donizetti, Frederick Delius, Bedrich Smetana, Randolph Churchill (the father of Winston), and of course Al Capone. But the disease does not preclude remarkable accomplishments. Hernán Cortés acquired syphilis in Hispaniola, but recovered and went on to conquer the Aztec empire.

The other major route of transfer for syphilis is a congenital one. The term congenital, which must be distinguished from genetic, refers to events that happen during foetal development and that are first seen at the time of birth – genes may contribute to a congenital defect, but the immediate cause is usually an external one. The congenital form of syphilis, picked up by a foetus from its mother during pregnancy, is very serious, and results in stillbirth about half the time. If the baby survives, the disease can develop through childhood and adulthood, culminating at a relatively early age in all the grim manifestations of tertiary syphilis.

Because congenital syphilis is present from the moment of birth, it affects the way the child develops. Among many other things the teeth grow into highly abnormal shapes, and the septum of the nose collapses (this can happen in tertiary venereal syphilis as well). While abnormalities of the skull and long bones that are typical of syphilis can easily be confused with similar defects caused by the related disease yaws, the distorted teeth are absolutely characteristic of congenital syphilis. This symptom provides a most important clue to medical detectives, for it is only present in a population that is infected with sexually transmitted syphilis. Even if a human population has left nothing behind but skeletal remains, the presence of abnormal teeth in the skeletons of young children is thought to be proof positive that congenital, and therefore sexually transmitted, syphilis was present in that population.

Was the disease present in Europe before Columbus's voyage? Syphilis has been called the great masquerader, because its huge range of symptoms often mimics other diseases. None the less, a number of the severer symptoms of syphilis, like the dreadful gummatous lesions of the tertiary form of the disease, are so striking and so well-defined that it surpasses belief that doctors of ancient and medieval times would not have described them in detail.

Clear records of syphilis began to appear in abundance just at the time of Columbus's voyages, a fact that soon led his contemporaries and many later historians of disease to make the obvious connection. The first description of something that sounds very much like syphilis is given in a letter written to a medical friend in June 1495 by Nicolas Squillacio, a doctor practising in Barcelona. His description is remarkably accurate, and leaves no doubt of the nature of the disease, which he reports to be widespread in the city He also says: '[T he sickness does not last more than a year, although the skin remains covered in scars which show the areas it affected.' The early lesions of the syphilis of today typically heal without scarring, so there is no doubt that the disease he saw was more severe than present-day syphilis. But his remark on the duration of the disease implies that by that time t had been present in Barcelona for at least a year. The inhabitants of Barcelona thought it had come from nearby France.

Syphilis began to afflict the troops of Charles VIII of France following the French takeover of Naples in February 1495, and this is often taken incorrectly to be its first appearance in Europe. Many of Charles's soldiers were so incapacitated by the disease that they were unable to take part in the battle of Fornovo during their retreat from Italy the following spring. But assuming that Squillacio was right and the disease had been known for some time in Barcelona, it must have appeared there before the French arrived in Naples.

The disease was not confined to Barcelona and Naples. Even at this early date, there are indications that it was widespread in other parts of Italy, where it was (inevitably, since the French were invading at the time) called the French pox. It was soon reported from all regions of Europe, and was described in detail by some of the most famous doctors of the time, including Paracelsus and Brassavola. Its obvious symptoms, and the ease with which it could spread, suggest that it might have been more virulent than the syphilis of today, though this may have been simply because the population was more susceptible. It was often claimed that it could be quickly fatal, and certainly the skin eruptions of secondary syphilis that were described at the outset of the epidemic seemed to have been more horrific than similar symptoms today. A woodcut by Dürer that appeared in 1496 but might have been executed earlier shows a knight whose entire body is covered with pustules.

It may be that many of the fatalities that were reported were actually caused by other diseases. But there seems to be no doubt that the disease was spreading like wildfire throughout the susceptible populations of

FIG. 9-1 Dürer's woodcut of a knight
apparently suffering from secondary syphilis

Europe. Its impact was great enough to provoke a flurry of royal and papal decrees. In view of its sexual transmission it makes sense that the disease should rapidly have evolved to become milder. Victims covered with pustules were hardly desirable sex objects, and strains of the disease that caused such a violent set of symptoms would quickly have been selected against. But even at the outset, victims like Cortés could manage to lead active lives for many years.

Syphilis was given its name by the Italian physician and poet Girolamo Fracastoro, who in 1530 published a long and highly allegorical poem about the disease. Part of the poem deals with symptoms and treatment. But Fracastoro also blended together the writings of the historiographer Gonzalo Hernandez de Oviedo with a fable from Ovid. The result was a tale supposedly told by the natives of Hispaniola about a shepherd named Syphilus, who angered Apollo and was punished with the disease. This astonishing and historically worthless farrago has been reprinted numerous times, and it appealed greatly to the learned of those days. When Erasmus became the first to name the disease itself after Fracastoro's imaginary shepherd, the rest of the literate world soon followed suit.

Attempts by historians to show that syphilis was present in Europe before Columbus's voyages, through the examination of documents, have led to equivocal results. The descriptions of nasty poxes and venereal diseases from before that time tend to be vague, and might as easily be descriptions of gonorrhoea or a number of other diseases. Students of the disease such as Richmond Holcomb and Ellis Hudson have suggested that many early cases of syphilis were confused with leprosy in the days before the invention of printing, when medical knowledge was small in scope and limited to whatever manuscripts a doctor might have access to.

There is no indication of the typical lesions of syphilis on skeletons that have been dug up and examined from the graveyards of the leprosy hospices that sprang up by the thousand in the twelfth and thirteenth centuries. Signs of leprosy in these skeletons, however, are clear and unequivocal. On the other hand, there is testimony from three witnesses who, though writing some time after the events, make a direct link between the disease and Columbus's men, and indeed are the first to propose an American connection.

The first is Bartolomé de las Casas, a priest who arrived in Hispaniola in 1502. He questioned some members of the rapidly disappearing native tribes of Hispaniola, and was assured that the disease had been present among them before the arrival of Columbus.

The second is Gonzalo Fernandez de Oviedo y Valdez, who in 1513 was appointed superintendent of the gold and silver mines of the New World. Oviedo had been present at the court of Ferdinand and Isabella during Columbus's voyages, and may have caught syphilis himself by the time of his appointment. It has been suggested that one reason he sought the post was because he wanted to track down and profit from native methods for curing the illness. Indeed, one possible native remedy, the wood of the guiac or lignum vitae tree that grows in Puerto Rico, was used for the treatment of syphilis, though with little success, throughout the sixteenth century. Regardless of his motives, in his communications with the king of Spain Oviedo was firm in his conviction that Hispaniola was the source of the infection.

The third is the Spanish doctor Diaz de Isla, who recounted having seen and treated some of Columbus's sailors on their return (though this claim appeared only in a manuscript version of his 1539 book). He gave a confusing picture of the origin of the disease, asserting at one moment that it came from Hispaniola and the next that it had been known since ancient times.

While the idea of a New World origin gained immediate favour and was supported by those who were most closely involved with the events, contrary voices were soon heard. The arguments on both sides have seesawed endlessly ever since. Initially, the opposition had more of a religious than a scientific basis. The biblical tales of the afflictions of Job and of the beggar Lazarus were adduced as arguments that syphilis had been present in the Old World since ancient times.

Ellis Hudson and other more recent opponents of the Columbian hypothesis turned to logistical arguments, pointing out how unlikely it would have been for the disease to spread so rapidly from Spain to Italy and France if it had been introduced by a handful of diseased sailors or captive Indians. This argument ignores documented evidence of a similarly rapid spread of plagues such as the Black Death. Of course syphilis, as it currently presents itself, is not overwhelmingly infectious and would be very unlikely to have spread as quickly as the Black Death. But since documentary evidence seems to support such a rapid spread, and the symptoms of syphilis seem to have been much more striking then than now, perhaps the disease had such a capability when it first appeared.

Much effort on the part of advocates of a Columbian connection has been wasted in looking for a direct link between Columbus's sailors and the French occupation of Naples – did some of the sailors somehow make their way to Italy and join the French army as mercenaries? Did some of the female Indian captives turn to lives of prostitution? There is certainly no indication that either of these things happened. However, if the disease had spread so rapidly as to infect the inhabitants of Barcelona heavily within a year of Columbus's return from his first voyage, such a direct connection need hardly be supposed. Nor should the absence of such a direct connection rule out a New World origin.

More cogent evidence has now been found. Beginning in the early twentieth century, bones from pre-Columbian burials n the New World and ossuaries throughout Europe and North Africa were intensively examined for traces of syphilitic alterations. Careful scrutiny of the evidence found no certain pre-Columbian cases of syphilis in Europe, and the results from North Africa are equivocal (only two out of 25,000 Egyptian skeletons, for example, showed lesions that might have been syphilis-caused). One reasonably certain case from Iraq, dating to 50 BC, has been found, but it might easily have been due to a severe case of another disease related to syphilis called bejel, which until very recently was endemic in the Middle East.

Much clearer evidence for numerous cases of pre-Columbian syphilis or yaws (probably the latter) has been found in old skeletal remains from Australia and the South Pacific. Yaws was certainly present in these more tropical regions, and even managed to accompany the Polynesian migrants on their long voyages. And yet it never spread into Europe, perhaps prevented by the great deserts that separate Europe from the warm moist tropical regions where yaws can thrive.

Similar surveys of skeletal material have been carried out in the New World, and here as in the South Pacific the contrast with Europe is striking. Many pre-Columban skulls found in North, Central and South America show a pattern of erosion, particularly of the frontal bone, that is characteristic of the tertiary stage of the disease. It is impossible to rule out the possibility that these lesions were caused by yaws rather than syphilis, though some of the discoveries were found on bones from burials in the temperate zones of both the Americas. As we will see, however, syphilis-like diseases can spread in the temperate zones by means other than sexual transmission.

Perhaps the strongest case for the existence of pre-Columbian venereal syphilis comes from the skeleton of a six-year-old child, found in Virginia and now preserved in the Smithsonian Institution. Abnormal new bone growth, typical of syphilis, appears in many parts of the skeleton, but the most telling evidence is the malformed teeth. There is no doubt whatsoever that this child suffered from congenital syphilis, and that therefore sexually transmitted syphilis must have been present among the pre-Columbian Indians of Virginia. No such deformed teeth have ever been found in pre-Columbian Europe, though the discovery of a single one would be quite enough to undo the Columbian theory.

So, after all this, have the advocates of a New World origin for syphilis won their half-millennium battle? Until recently it would have seemed so – a review of the evidence in 1988 that reached this conclusion was received by physicians, anthropologists and archaeologists with the kind of broad-based agreement that is almost never achieved in the highly fractious world of academe. Unfortunately for the Columbian hypothesis, however, in late 1994 a thousand-year-old Viking skull that showed apparently clear signs of syphilitic lesions was found 200 kilometres north of Stockholm. The story is anything but settled, particularly if we try to push the history of the disease back further than 1492.

Let us assume that Columbus did indeed bring the disease back with him from the New World, in partial payment for all the loathsome pestilences that he and the other European explorers were carrying in the opposite direction. The presence of syphilis on the Caribbean island of Hispaniola suggests that either it came with the peoples who crossed to the New World via the land bridge connecting Alaska with Asia, some twelve to fifteen thousand years ago, or that these peoples found the disease waiting for them. If the latter, then presumably the disease lurked in an animal host native to the Americas. Now, no animal host is known for the specific disease of syphilis, though yaws has been found to infect baboons and their relatives in Africa. Was there in fact an animal host in the Americas?

Much interest was generated a few years ago with the discovery of an apparently syphilis-like or yaws-like infection that had distorted the skeleton of a bear dug up in Indiana. The bones were dated to roughly

11,500 years ago. Perhaps, it was suggested, the newly arriving peoples in the area managed to pick the disease up from bears – we will refrain from speculating exactly how. But no signs of such a disease have been reported from living bears, making this scenario highly unlikely.

If on the other hand the first human migrants to the New World brought the disease with them, it must have been present in the Asiatic populations from which they were derived. And yet there are no signs of syphilitic infection in old Asian skeletons, and the more recent inhabitants of Asia seem to have been as susceptible to syphilis as the people of sixteenth-century Europe. Excellent historical records show that syphilis hit Japan like a thunderbolt in 1512 and soon became very widespread, with as many as half the population of Edo (Tokyo) affected during the seventeenth century. For the Columbian hypothesis to be correct, we must conclude that a particularly virulent form of syphilis arose in the New World and nowhere else, and came to Europe on Columbus's ships. The evidence for this is still anything but conclusive, and in any case the Columbian hypothesis begs the question of the ultimate origin of the disease.

A thread of merest gossamer

To trace the origin of syphilis we must move beyond history and archaeology to the worlds of medicine and molecules. A good place to begin is with the causative agent itself.

The bacterium of syphilis proved remarkably hard to find. It might have been glimpsed as early as 1837, with the discovery by an early microscopist named Donné of a 'spiral microbe' that could be seen in syphilitic lesions. But his observation was swept aside throughout the rest of the nineteenth century by a confusing multiplicity of discoveries of all sorts of bacteria in syphilitic lesions – all in retrospect probably due to secondary infections. Indeed, in view of the great difficulty microscopists had in finding the real culprit, Donné's spiral bacillus might also have been a secondary invader. Finally, in 1905, Fritz Schaudinn (who you will recall led the world badly astray in the matter of the malaria parasite) redeemed himself by tracking down the real culprit, although it was some years before his discovery was fully accepted and by then it was too late for Schaudinn – he died of

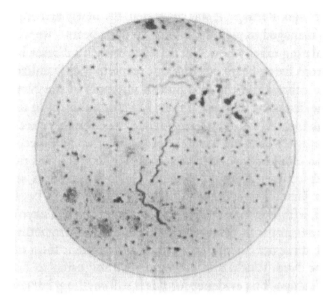

FIG. 9-2 The syphilis spirochete

septicaemia a year after his discovery. The organism responsible was a pale, almost transparent bacterium with a body twisted into a spiral, one of a large group of such spiral bacteria that have been gathered together under the name Spirochetes (Figure 9-2). It was given the name *Treponema pallidum*, which means pale twisted thread.

The difficulty in seeing it under the microscope is just one measure of this shy bacterium's fragility. *Treponema pallidum* is much smaller than its free-living relatives, and it is exceedingly feeble. A *Treponema* takes thirty hours to divide once, by which time an *E. coli* given unlimited resources would have produced 1,000,000,000,000,000,000,000,000,000,000 offspring. It is poisoned by too much oxygen, and it is virtually impossible to persuade it to grow in the laboratory even in a culture of human cells. Limited success has been achieved by the addition of chemicals that remove free oxygen, but even with a great deal of care *T. pallidum* invariably dies or stops growing after only a few divisions.

This stands in vivid contrast to its many relatives, the free-living spirochetes, some of which inhabit various places in our bodies – skin, mucous membranes, gut, and the gumline around our teeth. These spirochetes, far more robust, can be grown in artificial media, though they too multiply very slowly.

T. pallidum seems only able to survive under the narrow and highly specific set of conditions found in the human body. A small rise in temperature will kill it, as Wagner-Jauregg found when he introduced the successful malaria therapy for syphilis in the 1920s. (Fascinatingly, the fact that sufferers from high fever could be cured of syphilis had been noted by Ruy Diaz de Isla in 1539, which shows what remarkably keen observers the great doctors of the past could be.) And cold is fatal to it as well – when blood contaminated with *Treponema* is placed in the refrigerator, the bacteria die off so completely that after two or three days the blood is actually safe to use for transfusions. A bit of soap will kill it, too.

One might think that there is nothing to fear from such a nebbish of a parasite. Paradoxically, its very feebleness seems to enable it to survive in the hostile environment of its host. It is skilled at making do with very little, and its sheer doggedness makes it one of the most infectious organisms known. At the same time, like other harmful parasites, it is treading a fine line between survival and destruction.

Some of this feebleness can be traced to the fact that *T. pallidum* has very few genes. Its chromosome, like those of other bacteria, is a circular piece of DNA. The circle is made up of about a million base pairs, which may seem like a lot, but it is only a third the size of the chromosomes carried by more robust and independent bacteria like *E. coli*. The paucity of genes in its genome may help to explain two odd facts about it, one that contributes to its destruction and the other to its survival.

First, and most unusually, mutant strains of *T. pallidum* that are resistant to penicillin have never arisen. Because it is unable to exchange genetic information with other bacteria, it cannot acquire genes from them that would enable it to destroy penicillin, and it does not seem to have the genetic resources to make an enzyme on its own that can break down the antibiotic. This fact has permitted doctors to continue to use penicillin as the weapon of choice for curing syphilis. The antibiotic is as effective now as it was when it was first introduced half a century ago. But this defect was of course irrelevant to the survival of *T. pallidum* in the pre-antibiotic era, since it never normally encountered penicillin.

Second, because *T. pallidum* has so few genes, it manufactures only the merest excuse for an outer cell membrane. This threadbare coat is easily breached by even small changes in the bacterium's environment, which is an obvious disadvantage. But at the same time it contains few proteins that can alert the host's immune system. Flimsy though it is, its membrane none the less has the capacity to attract and bind proteins found in the host's blood. These proteins form an additional protective layer, and because they are invisible to the host's immune system they also help to conceal the bacterium from host antibodies and patrolling white cells. Such relative invisibility undoubtedly helps to explain why the bacterium can survive inside its host's body for years.

The bacterium seems to be adapted to life in humans, and only in humans – most other animals may break out with early lesions when deliberately infected, but the disease progresses no further. Surely then, this fragile web of adaptations implies that *Treponema* has had a long history of association with humans. Indeed, this association appears to be real – but not, it appears, as the agent of syphilis.

A family of treponemes

A famous crescent of fertile lands, with its horns pointing downward, bestrides the Middle East. This Fertile Crescent sweeps around northern Syria and western Iraq, extending from Damascus near the Mediterranean coast through Aleppo, Mosul and Baghdad. The eastern part of the crescent encompasses the valleys of the Tigris and Euphrates.

On the Euphrates, south-east of Aleppo, is the town of Deir-ez-Zor. In the early part of this century Deir-ez-Zor was a dusty place of mud-walled buildings, housing small communities of Catholic, Orthodox and Armenian Christians and a nascent middle class of Muslim shopkeepers and bureaucrats. Surrounding the town, and stretching along the fertile banks of the river, were small communities of Bedouin tribespeople who had recently abandoned their nomadic life to take up subsistence farming.

This was the situation in 1924, when an American clinic and hospital was established in the town. In the years up until World War Two the doctors in the clinic treated the townspeople and the Bedouin for a wide variety of diseases. Ellis Hudson, one of the doctors, has left a vivid record of what they found.

Venereal syphilis, which was commonly called *fraiji*, meaning foreigner, was very common among the Christians and less so among the Muslims of the town. It presented the full range of ghastly symptoms, including insanity and paralysis, typical of tertiary syphilis before the advent of penicillin. But syphilis was almost unknown among the Bedouin, and rare among the Muslim townspeople. They suffered instead from a disease called *bejel* that showed some of the same symptoms as syphilis. Bejel affected children as well as adults and had none of the moral stigma of syphilis – it was simply another childhood disease among many.

About seventy per cent of the tribespeople who came to the clinic, for whatever reason, had bejel. The external symptoms were sometimes as severe as those of tertiary syphilis, with bowing of the tibia, collapsed nasal bridges, and extensive skin lesions on many parts of the body. As with syphilis, lesions were common outside and inside the mouth. Only rarely, however, did they appear on the genitals, and there were virtually no cases among the Bedouin of the paralysis and insanity seen with tertiary syphilis.

The Bedouin were stoic about the disease. They suffered through the inadequate efforts of the doctors in the clinic, which consisted primarily of treatment with bismuth. And often they would vanish partway through their course of treatment, to the frustration of the doctors.

Syphilis, and probably bejel, have been common in North Africa and the Middle East for centuries, and Arabic medicine found dramatic ways to deal with them. One common method was to cover the head with a cloth and breathe the fumes of heated cinnabar, a mineral consisting primarily of mercuric sulphide. The fumes were essentially pure mercury vapour, which was driven off from the mineral when it was heated. This treatment produced, in addition to who knows what horrific brain damage, copious salivation. With luck the lesions, particularly the most bothersome ones around the mouth, would disappear.

Bejel was an annoyance, not a fatal disease, and while it could cause gross distortions of the body later in life, the hardships of Bedouin existence ensured that few people, with or without bejel, survived into old age in any case. This may have been why it did not give rise to the same neurological signs as tertiary syphilis – it simply did not have the time.

The Bedouin adults, their children, and their animals all lived in close proximity in their tents. They were afflicted, particularly during the brief spring, with fierce visitations of biting flies. And, even though the Euphrates flowed nearby, they followed the traditions of their earlier desert existence and never washed themselves and rarely washed their clothes. Soap, of course, was beyond their means. This accumulating filth provided ideal conditions for the transmission of bejel. Children living in insanitary surroundings are infected by contact with each other, by wearing each other's clothing, through shared drinking vessels, and perhaps through the agency of houseflies or blowflies which pick up the treponemes on their mouth parts and transfer them from an open wound on one host to a scratch or an open wound on another.

It seems likely that the infectious agent of bejel is slightly tougher than that of syphilis, for it can survive at least short periods outside the body. This ability opens up many more routes through which it can be transferred.

The history of European syphilis, as we have seen, can be traced back half a millennium with a moderate amount of certainty. The history of bejel, on the other hand, is essentially unknown. But we know that the disease, or something like it, must be far older than that. Hudson thought that both bejel and syphilis were old, that syphilis might have arisen from bejel, and that Arabic and early European physicians had confused matters by lumping the two diseases together with leprosy. He pointed out that mercury had commonly been used as an attempted cure for leprosy, and he cited repeated statements by physicians through the centuries that leprosy was passed by sexual contact.

But the symptoms of leprosy are clearly different from those of bejel. While there are indeed lesions on the skin, they have a very different appearance from either syphilis or bejel. The bone loss in the extremities that is so typical of leprosy is never seen in even the most advanced cases of these treponemal diseases. And unlike syphilis, there are plentiful examples from old burials of the highly diagnostic lesions that are caused by leprosy. The earliest of these have been found in skeletons dating from the second century BC that were excavated in the Dakhleh oasis of the Sinai. As we saw earlier, enormous numbers of skeletons exhibiting the typical lesions of leprosy have been found in Europe. Most of them date from the twelfth and thirteenth centuries, when the disease reached its peak. Had syphilis or bejel left such copious traces behind, there would be no argument about their presence in the ancient world, and the proponents of a Columbian origin would long since have retired from the field in defeat.

It has been suspected for some time that the treponemes of bejel and of syphilis are very closely related, perhaps more closely than either is related to the treponeme of yaws. In experiments carried out in the 1950s, however, it proved impossible to distinguish the treponemes of yaws, syphilis and bejel on the basis of antibody response in rabbits – the rabbits' immune systems treated all three as indistinguishable.

Further, it is not at all obvious how old bejel might be. There is nothing in the evidence to suggest that it might not be a relatively young disease. Indeed, it was not formally described by Western doctors until the 1920s. Could it perhaps have arisen *from* European syphilis, rather than, as Hudson thought, giving rise to it? We have no written record of the history of this or any other disease among the nomadic peoples of the Sahara and the Middle East, so it is possible to speculate quite freely about when bejel first appeared. The plentiful ancient skeletal remains of North Africa and the Middle East do not show the signs of bejel that its high rate of prevalence among more modern Middle Eastern populations might lead us to expect. None the less, though he had no real evidence, Hudson thought that bejel was common in the ancient world, had spread from Africa to Europe, and had given rise to syphilis.

Of course, the chief reason that the possibility of a recent origin for bejel has not been considered is that there appears to be a much more likely source for it in the vast pool of yaws lying to the south of the Sahara. The scenario that Hudson and many others embraced is that bejel arose from yaws at the edge of sub-Saharan Africa, when strains of that treponeme appeared which were able to live in the deserts of northern Africa and the Middle East. Then, they supposed, the treponeme went a step further and acquired the ability to be transmitted sexually, which permitted its spread into the colder climes of northern Europe. Thus there might have been no Columbian connection at all, but simply a coincidence of timing brought about by the fact that Europeans began to invade both Africa and the New World at about the same time. They would have brought syphilis to Europe from its more obvious source, Africa.

Unfortunately for this theory, there had been many contacts between the civilized centres of the Middle East and Africa, even sub-Saharan Africa, since the time of the Phoenicians and probably before. Why didn't yaws or bejel make the jump to syphilis, and spread to the Middle East and beyond, during those three millennia when the Phoenicians, the Arabs, and even the Chinese explored the eastern coast of Africa?

Much less is known about yaws than about syphilis. Even the origin of the word yaws is uncertain. It may be African, brought to the New World along with the slave trade. Or it may come from a Carib Indian word that simply means disease. Like bejel, yaws is a highly communicable skin disease, though it is found throughout the humid tropics rather than the dry deserts. Its symptoms, like those of bejel, do resemble those of syphilis, but just as in bejel there is normally no involvement of the central nervous system.

Yaws is found now throughout the humid tropical world, but it must have arisen somewhere, and the best guess is Africa, since sub-Saharan Africa remains the major focus of the disease. It probably accompanied the ancestors of our species on their first migrations into other parts of the Old World sometime between two hundred thousand and two million years ago, and we certainly know that it accompanied the Polynesians who spread out over the eastern Pacific over the last two millennia.

Yaws seldom kills, but it can be debilitating and ultimately devastating. Like syphilis, it is easily cured with penicillin, and in the 1950s and 1960s the World Health Organization gave penicillin to fifty million people with the disease. The impact of the antibiotic was initially striking, but as the incidence of yaws diminished and it became necessary to go to more and more remote areas in order to track down the remaining cases, the interest of the governments involved waned.

The result has been a recent dramatic resurgence of infection, particularly in Africa during the 1980s, one of the many grim results of the recent breakdown in social structure in that tormented continent. A continued focus of infection has been among the Pygmies of the Central African Republic. These shy people are constantly on the move and difficult to reach with medical care.

Yaws, bejel and syphilis, however, are not the only diseases on this continuum. A rare, very mild yaws-like disease called pinta is found in Central America and the rainforest side of the Andes. And a bejel-like disease, also not transmitted sexually, occurs in the Balkans and as far north as European Russia. These latter areas are very different from the dry hot deserts where bejel is usually found, or the humid tropics that are the abode of yaws.

The Balkan form of the disease is transmitted in an unusual way. As in the other treponemal diseases, lesions are commonly found inside the mouth. Mothers in these regions often chew solid food to make it easy for babies to swallow, and this habit enables the treponemes to spread. How inventive these treponemes are at finding bits of the warm humid tropics in the temperate zones! And how surprising, if the Columbian connection is true, that they did not, in spite of all their opportunities, manage to invent the venereal mode of transmission in the Old World!

Yaws, bejel and syphilis are caused by spirochetes that are all so similar to each other that they have been lumped together under the species designation *Treponema pallidum*. Astonishingly, there is still no certain way to distinguish among them. They were originally given different species names, *pallidum* for the syphilis and bejel spirochete, and *pertenue* for the yaws spirochete. But all three look exactly alike, even under the electron microscope, none of them can be cultured, and all stimulate experimental animals to make what appear to be the same set of *Treponema*-specific antibodies. This means that they cannot be distinguished by immunological tests. *Pallidum* and *pertenue* have now been demoted to subspecies of *T. pallidum* by microbiologists who know that they have to be different but are unable to find the differences.

Even sophisticated molecular methods have not turned up consistent differences. Several genes have been examined in the two subspecies, and are identical right down to the DNA level. At one point hopes were raised with the finding, in one enormous protein that is made up of almost 20,000 amino acids strung together, of a single amino acid difference between the two subspecies. These hopes were dashed again with the immediate discovery that in each of the two subspecies some strains make one type of the protein and other strains carry the other type.

Of course there have to be some differences somewhere in those million or so bases of the *Treponema* genome, and 'libraries' of *Treponema* DNA are now being scanned to try to pin down some of them. Whatever and wherever the differences may be, some of them must explain why the strains that cause yaws and bejel are more likely to be transmitted by non-sexual contact than the one that causes syphilis, and why the syphilis bacterium is more likely to produce the internal and external genital lesions that allow it to be transmitted sexually. And finally, they must explain why the syphilis bacterium is able to invade the dorsal spinal cord and the brain, while the others do not – or do so only rarely.

Might it be possible that the differences among these closely-related bacteria are the result of a kind of evolutionary trade-off? As the treponemes evolved a greater ability to invade the body of their hosts, they may at the same time have lost their ability to survive elsewhere. The delicate structure of their outer membranes is the result of selection for strains that present the smallest possible number of proteins and carbohydrates to their hosts' immune systems. So delicate have the membranes become that it appears that they now need to be reinforced by proteins borrowed from the host. The bacterium has become a ghostly little creature. By the time its host's immune system perceives and reacts against it, it has penetrated deeply into parts of the body where it inadvertently does lasting harm.

This Faustian bargain has had its advantages. If one plots the incidence of these diseases out on a map, it is obvious that syphilis is the one that has been most successful in spreading to the world's temperate zones. It was this distribution that led Cecil Hackett, a senior medical officer of the World Health Organization in Geneva, to propose in the 1960s yet another scenario. He suggested that pinta was once widespread throughout the world, and was replaced by yaws in Africa. Then yaws gave rise to bejel, which in turn gave rise to venereal syphilis as people began to wear more clothing and to live in towns and cities. All three of the newer diseases were shaped as much by the environment as by genetic differences, and presumably each could give rise to the others. As a consequence, he dismissed the Columbian hypothesis.

The universal use of penicillin has been an enormous blessing, but has also destroyed most of the evidence about how these strains might have been related to each other, and how easily one might have evolved into another. Mutations in the treponemes that increase or decrease their virulence do seem to be common, and there is some evidence that such mutations take place at the present time. There are some controversial indications that in Papua New Guinea and elsewhere, a new type of yaws is becoming more frequent. It has been called attenuated receding yaws, and is even milder than pinta in its symptoms. This mild form seems to have become more prevalent since the widespread use of penicillin, and its apparent lack of severity may have contributed to the loss of governmental interest in tracking down and eradicating the remaining cases of yaws.

Occasionally the yaws treponeme can mutate in the other direction. In 1990 a little Indonesian boy was discovered who had syphilis-like ulcers on his genitals as well as the rest of his body. The case was first diagnosed as venereal syphilis, but on closer examination it seems to have been a particularly syphilis-like case of yaws. It remains to be discovered how many of these changes are due to genetic alterations in the treponemes and how many are due to changes in the environment.

There is something terribly unsatisfactory about any evolutionary scenario that can be constructed from our present knowledge. If the ancestor of yaws or pinta first infected our ancestors in Africa, then presumably, as our species spread through the warmer parts of the Old World, we must have carried the spirochetes of yaws with us from our ancestral home. But for some reason these spirochetes did not immediately adapt to drier and colder climates by the obvious step of selection for mutant strains that could be transmitted sexually. The Columbian scenario compels us to suppose that they somehow managed to survive the hundreds of millennia of human spread across Asia, and the final wintry transit across the Bering land bridge during the last ice age. Then, once they reached the New World tropics, they belatedly, perhaps in the Caribbean and in a few other places, mutated to strains that could be venereally transmitted And it was these strains, remote descendants of their African forebears, that Columbus and his sailors brought back to the susceptible populations of the Old World.

For this scenario to work even approximately, it has to be assumed that venereal syphilis was confined to fairly small regions in the New World, since Indian populations in places other than the Caribbean were devastated by syphilis when it was introduced (or reintroduced) by Europeans.

Surely though, one might think, if the treponemes are all so closely related, then at some point during all those hundreds of thousands of years as our ancestors huddled together on the windy steppes of Asia, they would have managed to mutate to a venereal mode of transmission.

Even if the recent out-of-Africa theory of human origins is correct, and modern or near-modern humans arose in Africa and spread from there through Asia as recently as one or two hundred thousand years ago, this is still a very long time for the treponemes to refrain from taking this obvious evolutionary step.

Alternatively, while yaws certainly arose in the Old World (or else it could not have been spread from Asia throughout the South Pacific by the Polynesian voyagers), pinta and syphilis could have been acquired by our species in the New World, perhaps by transfer from Pleistocene bears or some other animal source. Arguing against this possibility is the fact that a really likely animal source for the treponemes has not yet been found in the Americas. And it seems very implausible that a treponeme genetically indistinguishable from that of yaws, already well adapted to survival in humans, could have somehow evolved in other animals in the Americas before humans arrived.

This of course is the essential frustration faced by the many epidemiological detectives who have tried for centuries to track down the source of syphilis. Whether we assume a New World source or not, we must assume long periods during which the spirochete did not take the evolutionarily obvious step of moving to sexual transmission. And it is at this point that the arguments usually tail off, leaving us unable to decide among all the competing possibilities.

Again and again in the course of this book, we have seen how the spread of epidemic diseases represents – in figurative terms – an act of desperation on the part of the organisms that cause them. When old avenues are closed to them they must find new and effective routes by which to spread through their host population, but often only by putting themselves in harm's way at the same time. It appears that the treponeme of syphilis is no exception. It has repeatedly been observed that venereal infection is an excellent way for the treponemes to get around the problem of a cold dry climate where people wear clothes, the skins of children rarely touch, and the only really intimate contact is sexual. Ellis Hudson, working in his Syrian village, observed that the Muslim people of town and countryside had far fewer indiscriminate sexual contacts than the Christians. He proposed this to be the explanation for the prevalence of syphilis among the Christians and of bejel among the Muslims. Perhaps, as he and many others have suggested, it is the rise of civilization, and of the more frequent sexual contacts among people living in the wicked metropolises that resulted, that has caused the spread of syphilis.

This argument seems a mite strait-laced and forced. After all, fairly indiscriminate sexual contact is not unknown in many native cultures. But perhaps the argument can be extended a bit. The treponemes have only a limited number of options open to them. As we saw, there are many culturable relatives of *T. pallidum* that live on the skin, possess much larger chromosomes and thus many more genes, and yet do not produce disease. These culturable treponemes, though still picky about their growth conditions, also multiply more quickly than *T. pallidum*. In order to penetrate the body even partway, it appears that *T. pallidum* has had to shed genes in a kind of genetic striptease, slowing its growth but at the same time getting rid of proteins that might alert the host's immune system. Only that ultimate bacterial ecdysiast *T. pallidum pallidum* seems capable of working its way sufficiently far through the body's defences to take up a primarily venereal mode of transmission.

Thus, the germ of syphilis absolutely depends on a relatively mobile host population with a high frequency of indiscriminate sexual contact, situations that may have existed locally in the ancient world but which probably did not become the norm until cities grew in size and travel became easier. Remove these conditions, and syphilis should disappear or fall to very low levels. Perhaps syphilis arose locally and died out many times, in both the Old World and the New, leaving no more than a few equivocal traces among the skeletal remains of the ancient world. Perhaps Columbus and his men stumbled on and brought back with them a particularly virulent strain from Hispaniola, or such a strain was brought in from Africa by the Portuguese navigators. Alternatively, a sexually transmitted strain might have arisen spontaneously in Europe around that time, and coincidentally happened to spread because travel was becoming common and cities were growing crowded. Then Columbus's voyage could be seen as a kind of symptom of the disease, and not its cause!

It should soon be possible to make choices among these various alternative histories. If I am right, the syphilis treponeme should be separated from its close relatives, not by some new gene that confers on it the ability to be transmitted sexually, but by the loss or damage of a gene or genes. These will likely prove to be genes making proteins that confer on the treponeme's outer membrane some of the integrity needed for it to survive more effectively in the outside world, but that at the same time make it visible to the host's immune system. The bacterium that carries such mutations has gained the ability to penetrate and form damaging lesions deep within the human host, but at a cost, for now its easiest way to escape to a new host is through sexual transmission. And, because damage to already existing genes is much more likely to happen than the appearance of a new gene, then syphilis could have arisen many times. But one or more of these mutant strains only became widespread when conditions were favourable, as they apparently were during the Age of Exploration. This would make the whole Columbian connection business rather moot.

More information is obviously required. And some may have been supplied by a recent and fascinating discovery that was made using highly specific monoclonal antibodies to *Treponema pallidum*. George Riviere and his colleagues, working at the Oregon Health Sciences University and other institutions, investigated the bacterial populations of patients with severe periodontitis and necrotizing ulcerative gingivitis. (Your dentist may have shown you pictures of these diseases, in order to terrify you into brushing and flossing faithfully.) Many of the bacteria from such infections cannot be cultured successfully, but their presence can sometimes be detected by antibodies. When certain antibodies specific to *Treponema pallidum* were used, results were negative for 24 uninfected subjects, but about half the patients with each of the gum diseases tested positive. These people were, unsuspectingly, carrying around a microbe that seems to be very similar to the microbe of syphilis – though whether it actually caused the gum disease remains to be discovered.

This casts an entirely new light on the question of the origin of syphilis. The bacterium and its allies may be far more ubiquitous than we thought, and may have found another completely different way to escape from the tropics. If *T. pallidum*, or something very closely allied to it, can really be carried around as a member of the bacterial population in our mouths, this might help to solve the vexing question of how it could have made its way across the icy roof of the world to the Americas. It might have survived all that time by being transferred from the mouths of mothers to those of their children, perhaps in the form of the bejel-like disease of the Balkans and European Russia. Of course, we are still faced with the problem of why it then evolved to take up a sexual mode of transmission in the Caribbean. Perhaps the mother-to-child mode of transmission was blocked by changing habits when the New World migrants reached the tropics

We may never know all the details of this fascinating and frustrating story. Yet syphilis illustrates vividly how difficult it is for a pathogen to escape from the tropics to the colder, less hospitable world of the temperate zones, and how dependent it can become on the specific behaviours of its hosts. The syphilis treponeme has made an evolutionary Faustian bargain, a bargain that has opened up the world to it – but at a price.

So far as these creatures are concerned, the human sex act is merely a little bit of the warm, moist tropics that has been transferred to the temperate zones. To take advantage of this opportunity, however, they must, unlike their tropical relatives, be very resourceful at overcoming the immunological defences that our bodies have thrown up. It seems hard to imagine that if similarly feeble bacteria were not transmitted sexually they would have made much headway in the temperate zones.

Syphilis is of course not alone in being assisted in its spread beyond the tropics by secrecy, embarrassment and prejudice. AIDS has made a similar escape, and is a far more dangerous foe, for it does not succumb tamely to penicillin. Yet the AIDS virus, despite its fearsome aspects, has had just as much difficulty in spreading through the human population as syphilis or typhoid, and has had to make equally dramatic compromises in order to retain its ability to spread. AIDS the greatest current epidemiological threat to our species, makes a fitting introduction to the large themes of the final chapters – the role that diseases have played in shaping the world we live in.

PART FIVE

Plagues, Populations and the Biosphere

10

AIDS and the future of plagues

Is AIDS another legacy of African colonialism?

In Africa, unlike Western Europe, borders mean something. They separate entire psychological mind-sets. A few years ago, a group of white South African soldiers guarding the border with Swaziland treated me with contempt as I ventured across it, for being stupid enough to foray into the heart of darkness. (Swaziland is actually a beautiful and well-run little country which is making great strides in conservation and education with the aid of gambling dollars from the nearby South African city of Durban.) And when I later stepped across the border from Zimbabwe into Zambia at Victoria Falls, I was transported from a lively tourist scene into a land of depressing discouragement. Nobody in Zambia seemed to have the energy to figure out how to rake in some of the tourist money flowing into Zimbabwe a few steps away.

The tourist sees nothing, however, of the true weight of misery that contributes to this sense of discouragement and hopelessness. Throughout this century, sub-Saharan Africa has been marked by dramatic population growth, punctuated by immense slaughters and by vast movements of whole peoples. The resulting destruction of traditional ways of life has caused repeated outbreaks of disease.

Uganda is one of the countries hardest hit by the current AIDS epidemic, but this is only the latest in a series of disasters over the years. In 1907, an outbreak of sleeping sickness killed half the population. More recently, during the 1970s and 1980s, two of the outstanding thugs of the twentieth century, Idi Amin and Milton Obote, directed the killing of a total of at least 800,000 people who happened to belong to tribes different from themselves. A million refugees were forced to flee the country. Unsurprisingly, much of the society of Uganda, once called the Pearl of Africa, has been destroyed.

While the Plague of Justinian and the Black Death had effects on the size of the population of Europe that lingered for centuries, more recent outbreaks of disease and warfare have had surprisingly little effect on the growth of human populations. People have always been resourceful enough to find ways to go on living and reproducing under the most frightful and dangerous conditions, and now instant communication can in most cases alert relief organizations to the spread of starvation.

Thus, in spite of its grim recent history, the population of Uganda has continued to grow. But the AIDS epidemic seems likely to kill a sizable fraction of this much larger Ugandan population during the last decade of this millennium and the first decade of the next. A similar pattern of intertribal conflict, disease and massive displacement of the population is currently taking place in Rwanda, where the spectre of AIDS looms even larger.

Will AIDS be devastating enough to have an impact on population growth in these countries? Pronouncements on this subject by Western scientists have tended to be less than helpful, and to recoil on those making them in unexpected ways. In June 1992, the epidemiologist Roy Anderson, who was then at the University of London and is now at Oxford, announced that he and ecologist Robert May had calculated that in a number of the countries of Central Africa the impact of AIDS will be horrific. In fifteen years, their populations will actually begin to shrink. Unfortunately, their prediction seems to have provided a further justification to the governments of some of these countries to discourage birth control and educational measures – for what is the point of embracing the politically embarrassing message of birth control if AIDS is going to do the controlling for you? And what is the point of spending money to educate people if so many of them are going to die anyway?

These demographic projections notwithstanding, I doubt that AIDS will have much of an effect on population growth.* Ignorance, war,

---

* The highest rate of human population growth ever recorded took place, not in Africa or in Asia, but in French Canada at the beginning of this century. Now, a hundred years later, and even though this group is predominantly Catholic, it has one of the lowest rates of growth. This astounding demographic transition has certainly not come about through the agency of disease.

famine and disease already stalk Africa today in a way that is reminiscent of Europe in the fifteenth century. Ten million Africans mostly children, die every year of dysentery and allied diseases, and another million die of malaria, yet the population continues to explode.

It will take a while to change the prevalent mind-set in Africa. Late one night, in a camp in the Okavango delta, I talked with a Botswanan who had fathered a dozen children by as many women. He exhibited a rather disarming mixture of pride and embarrassment as he recounted these exploits, and admitted that if he had known in his youth what he knows now, he might not have cut quite such a swathe through the ladies and fathered quite such a crowd of offspring. Why so many children? He was quite open about it. In a world in which he felt inferior, it was the only way in which he could prove his manhood.

Whether or not it halts population growth, there is no doubt that the impact of AIDS on African societies will be much greater in the future. But with the notable exceptions of Uganda and Zambia, most African countries have until recently denied that it is a problem, and have refused to establish educational and prevention programmes. My Botswanan companion's confession was unusual, for reticence about discussing sexual matters is a hallmark of many African societies. As a consequence it has been very difficult to control the spread of all sexually transmitted diseases, not just AIDS.

AIDS has stabbed at the very heart of the poorest countries in Africa and Asia, causing the potentially most productive members of these societies to be struck down in their middle years and paralysing these countries' already inadequate health care systems. There is no doubt that the disruption and fatalistic carelessness that is engendered by the high probability of dying from AIDS has contributed to the virtually complete societal breakdown in war-torn countries like Uganda, Rwanda and Somalia. It is difficult for most of us to imagine living in a country in which a large fraction of the population, very likely including ourselves, will be doomed to die young from a wasting disease. In such a society it is not possible to see a future.

There have been many epidemiological, demographic and economic studies of the impact of poverty and disease in sub-Saharan Africa. But what cannot be measured is the sheer weight of discouragement and the pervasive sense of inferiority that AIDS and all the other multiplying disasters inflict on the populations of these countries, and their effect on how sub-Saharan Africa is perceived by the rest of the world. Africans can hardly help it if they have been born in the region of the tropics that happens to harbour the largest number of endemic human diseases. None the less, there is an undercurrent of fear and contempt that poisons the relationship of Africa to the rest of the world. Most First World governments and corporations have cut back dramatically on their aid and investment in recent years. In 1994, the US Agency for International Development closed eight out of its 35 missions in sub-Saharan Africa.

People or cattle?

Twenty million people died during the great influenza epidemic of 1919. Perhaps twice that many died in Europe alone during the Black Death of 1348, with an additional unknown number of victims in Asia. But the AIDS epidemic, agonizingly drawn out over a span of decades, will dwarf any epidemic that has gone before it.

The World Health Organization projects that, by the year 2000, between 40 and 100 million people, chiefly in Africa and Asia, will have become infected with the two closely-related viruses that cause AIDS, HIV-1 and HIV-2. By that time, again chiefly in Africa and Asia, there will be between two and four million new cases of HIV infection each year.

Ten million of this total will be children. About half of them will acquire the virus from their mothers at the time of birth, but the other half will have contracted it through blood transfusions. Doctors and nurses who are trying to treat the severe childhood anaemias caused by sickle cell disease and malaria are forced to use improperly sterilized needles and unscreened blood supplies. This problem is found throughout the Third World, where nurses are forced to scrub disposable plastic syringes with soap and use them again. Soon, in Africa, AIDS will become the second leading cause of childhood death, passing the toll due to malaria – though still well behind the frightful carnage caused by diarrhoeal diseases. By the year 2010, current projections are that AIDS will account for half the deaths in sub-Saharan Africa. If the projections are correct, this will more than undo all the advances that have been made against other diseases in recent years, hard-won advances that are the result of public health measures and vaccination.

In First World societies, on the other hand, AIDS has until recently been largely confined to homosexuals and to intravenous drug users and their children. If this should remain true, the incidence of new cases of the disease in the United States and western Europe will probably remain steady at about 80,000 a year for the rest of the decade. The rate of new cases among homosexuals has dropped because of an increase in precautions, though no such reduction has been seen among the extremely hard-to-reach intravenous drug users. All this presupposes, however, that nothing happens to the notoriously mutable virus over the next few years. It is easy to imagine nightmare scenarios in which mutant virus strains suddenly appear that can be transmitted by mosquitoes, or simply through the air. Or, more likely, a strain might arise that is easily transmitted through heterosexual intercourse even under the conditions that are found in the developed world. In either case, one terror facing epidemiologists and policymakers is that the current AIDS epidemic in the developed world might expand into something that affects a much larger segment of society, so that it will begin to resemble the pattern seen in Africa and Asia. This is unlikely, but its probability is not zero.

At least at the moment, the courses of AIDS in the developed and developing worlds follow very different trajectories. The differences are so great that names have been given to distinguish the two patterns of infection – Type 1 in the First World and Type 2 in the Third. Homosexual transmission and direct infection by contaminated needles predominates in Type 1, while heterosexual transmission is the overwhelming mode of transfer in Type 2. Thus, the nightmare of First World epidemiologists, that the AIDS virus will acquire the ability to spread heterosexually, has already come true in Africa.

Yet the HIV-1 viruses that trigger these two types of outcome are not themselves noticeably different. The differences between Type 1 and Type 2 infections must be due to other influences. What are they?

There are four major factors. First, it is not the AIDS virus itself that kills its victims, but rather a whole zoo of opportunistic diseases. These breach the HIV-infected patient's weakened immune system, and then run rampant. As we will see in a moment, the make-up of this zoo is clearly and startlingly different in the two types of epidemic. This is reflected in their different time courses. In the developed world the progression of AIDS is agonizingly slow, with death coming an average of ten years after the first infection with the virus. In the Third World, particularly in Africa, the disease progresses far more swiftly. Most of the virus's victims will begin to lose the protection of their immune systems as little as two years after infection, and will die after another year or two from the effects of multiplying opportunistic diseases.

The second factor is the merciless exploitation of Africa by the First World governments and corporations that have preyed upon that resource-rich and helpless continent. From the beginning, the Europeans reinforced, extended and transformed the traditional systems of slavery that they found, modifying them to their own ends. Colonial powers, particularly those occupying West Africa, used slavery within the continent itself as a tool of subjugation. In some parts of West Africa, as we saw in the Gambia, slavery lasted right down to the beginning of the twentieth century. And even when slavery was supposedly eradicated, the system that replaced it was often even worse. In 1885, the Berlin Congress gave King Leopold II of Belgium permission to set up the Congo Free State. The Congress had in mind the noble goal of eradicating slavery in the area. Instead, Leopold, one of history's most venal figures, replaced slavery with something worse, a system of forced labour in which entire villages were destroyed and the surviving men and women were separated and sent to different parts of the country.

The impact of slavery, and of the forced labour systems that replaced it, on the African population was immense, far greater than the impact that any disease has had since. Throughout most of sub-Saharan Africa, populations declined dramatically during the nineteenth century and even into the twentieth. As a result of forced labour in the rubber plantations, the population of the Congo fell from thirty million in 1890 to eight million in 1924.

The companies that built the railroads, opened the mines and cleared huge areas to establish plantations of coffee, rubber, cocoa and bananas, treated their African labourers not as people but as cattle. It was cheap to move men just once over the long distances from their villages to the mines and plantations, and then to jam them together in dormitories and shantytowns. It would have been far more expensive to move their families as well, for this would have required the establishment of company housing rather than barracks.

This economic imperative to reduce labour costs had an enormous physical and psychological impact. Native African societies were so fragile that they were already crumbling quickly. Since most of the Europeans had no understanding of the value that those rapidly vanishing and very different social structures had in stabilizing African societies, they had no compunction about destroying them.

The third factor in the generation of the AIDS epidemic was an alteration in the pattern of sexual contact. Women, many of them the abandoned wives of migrant workers, soon arrived at the camps. They managed to survive in a great variety of ways, but many had to become prostitutes. The result was the appearance of whole sub-sectors of society in which the interaction between men and women was entirely based on prostitution. A similar situation had developed during the exploitation of the American West, but only briefly, and it involved relatively small numbers of people. In the gold, diamond and copper producing regions of Africa, companies like De Beers and Union Minière de Haute Katanga created a new societal pattern based on family disruption and prostitution, a pattern that persisted for decades and involved millions of workers.

The situation is beginning to change. Zambia Consolidated Copper Mines, among others, has established company housing and schools, and its workers can now live a semblance of a normal life. But the damage has been done. AIDS and other sexually transmitted diseases are as rampant among the workers at this company's mines as they are in the rest of Africa's so-called copper belt.

The final factor in this grim pattern of the AIDS epidemic in Africa is the prevalence of other sexually transmitted diseases. Chlamydia and trichomonas infections, in particular, often cause genital eruptions and sores, facilitating the transmission of the AIDS virus.

Francis Plummer of the University of Manitoba has collected data strongly supporting a connection between these diseases and the spread of AIDS. Plummer and his colleagues followed the fate of 163 Nairobi prostitutes over a period of two years. None of them had shown signs of HIV infection at the outset of the study. At the end of the two-year period, about half of them had developed antibodies to HIV. He found that if they had developed genital ulcers or chlamydial infection during this period, this greatly increased the chance that they would acquire the virus. Chlamydial infections are especially likely to erode the tissues of the cervix. When the women in his sample became infected with

HIV, the virus was commonly found in abundance in these damaged tissues. Syphilis and gonorrhoea, on the other hand, seemed to have little effect. In women, the virus's passage appears to be aided only by certain types of infection, particularly those that breach the layer of mucus that protects internal tissues. Among infected men, however, an association has been found between gonorrhoea and HIV transmission.

In Africa, and more recently in Asia, societal upheavals, brutal economic exploitation, and the unchecked prevalence of diseases like chlamydia and trichomonas have resulted in a particularly lethal combination. A large part of the blame for the AIDS epidemic must be laid at the massive and brassbound doors of Western multinational corporations that have been allowed to operate without regulation by both colonial and post-colonial governments.

It is a lucky thing for the majority of people in the First World that conditions there have been very different, so far confining the spread of the virus to particular subgroups. There is no guarantee, of course, that this will not change.

How we think HIV works

Can we use the same detective-like approach that we have taken with various other diseases to dissect how the AIDS virus does its damage? And to what extent can we apply the lessons we have learned from those other diseases to this new and frightful slow-motion plague? The synergism of HIV with other disease organisms, coupled with societal upheaval, has aided its spread in the tropics – will we find that, like malaria and syphilis, AIDS is operating at the very limit of its capabilities as it spreads into the temperate zones?

Some bald facts to begin with:

AIDS, Acquired Immune Deficiency Syndrome, is, so far as we know, a disease unlike any other that has challenged our species. It can be triggered by two different but closely-related viruses, members of the group called retroviruses. These viruses are capable of invading certain cells in our bodies and then causing the destruction of essential parts of our immune system. Like all other viruses, they are minimal bits of life that at some point in the distant past must have broken free from the cells that gave them birth. They left behind them in the process most of the genetic and biochemical machinery that is involved in the cells' survival and reproduction. As a result, viruses inhabit a twilight world between life and non-life, and must periodically re-enter living cells in order to make copies of themselves.

The genetic material of the group of viruses called retroviruses, to which HIV belongs, is made up of RNA, ribonucleic acid rather than the more usual DNA. This makes them very different from our own cells. Our cells, in common with the cells of most other organisms, make DNA rather than RNA copies of the DNA in their chromosomes in the course of their reproduction. As a result, we tend to think of DNA as the master genetic material, the ultimate repository of information in a cell. The RNA molecules that are copied from this DNA are for the most part mere molecular slaves, carrying the DNA's information from the nucleus to the rest of the cell.

But in the submicroscopic world, other rules often apply. There are some viruses that use RNA exclusively as their genetic material. They employ special enzymes that can make RNA copies of their RNA directly, and they never make any DNA at all. This scheme works perfectly well for these viruses, since there can be just as much genetic information in RNA as in DNA. It is thought by many scientists that such viruses may provide a glimpse of a very ancient world at the outset of the evolution of life, before DNA appeared. In that remote time it was RNA and not DNA that was life's master molecule.

Retroviruses such as HIV and many other similar viruses are not as extreme in this regard as pure RNA viruses. During the periods that the tiny virus particles float free in the environment, away from living cells, all their genetic information is in the form of RNA. But this changes dramatically when retroviruses enter the cells of their host. Their RNA molecules carry, among other things, the information needed to make a remarkably versatile enzyme called reverse transcriptase. Some molecules of this enzyme are carried around in the virus particle, and others are made after the virus enters the host cell. There, the enzyme quickly makes a DNA copy of the viral chromosome. The discovery of this remarkable reversal of the usual flow of genetic information was made independently by the late Howard Temin and David Baltimore and led them to a Nobel Prize in 1975.

The DNA copy of the viral chromosome, with continued assistance from the reverse transcriptase, then goes a step further. It violates the very essence of its victim's being, by entering the nucleus of the cell and inserting itself into the host's chromosomes. Most retroviruses can only accomplish this feat when the host cell is dividing, and it is here that the AIDS viruses have a great advantage. They can actually insert copies of themselves into the chromosomes of non-dividing cells. This opens up to them whole classes of cells that are normally protected from retroviral invasion.

If this invasion had gone no further than the insertion of the viral genetic information into the host's chromosomes, it might not matter much. Indeed, our chromosomes already bear multiple signs that such invasions have occurred in the past, in the form of about ten thousand fragments of retrovirus-like DNA sequences that are scattered here and there along their length. These fragments show that the cells of our ancestors were repeatedly invaded by retroviruses.

The chromosomes in all the cells of our bodies carry these bits of old retrovirus. This means that we have inherited them from our parents, and that those ancient viruses actually managed to invade our genomes in an even more intimate manner than HIV. At some point they somehow got into the cells that make the sperm and eggs that are passed on to the next generation.

Because they then lost the ability to re-emerge and become free-living viruses again, these old retroviruses are now quite harmless. They have become a permanent part of our genetic heritage. It is impossible for us to tell now whether some of them once caused AIDS-like diseases, for over the many millions of years that they have resided in our chromosomes they have undergone many changes. Because they no longer have a function, they have become moth-eaten from the accumulation of tiny mutations, and have even been broken up into pieces as the genetic material within our chromosomes has shifted about. As a result, many of them are barely recognizable.

There is no evidence as yet that the AIDS viruses have ever managed to invade our sex cells in a similar fashion. If they can, then they should be able to incorporate themselves into the chromosomes that we pass on to our children, becoming a permanent part of the genetic make-up of our species. Some surveys to determine whether this has happened have been carried out, with clearly negative results. At the moment, at least, the AIDS viruses and their relatives seem to be confined to classes of cells in our bodies that are not passed on to the next generation.

Once they become resident in these cells, the genes of the AIDS virus lurk quietly, awaiting a signal to make copies of themselves. The signal might be an intimation that the cells themselves are ready to divide, perhaps as the result of stimulation by an antigen or by a chemical message from other nearby cells. When the signal is received by the viral DNA, it is copied back into viral RNA. The copies are then coated with viral protein, and these new virus particles bud in huge numbers from the surface of the cells, damaging them in the process.

At the very outset of infection, the AIDS viruses may cause a mild, flu-like set of symptoms in some of their victims. Large numbers of viruses are made, and it is during this brief period, long before the victim suspects infection, that they are most likely to be passed on to a new host. Then, soon after infection, whether the viruses make their victim sick or not, the host's immune system reacts and begins to manufacture antibodies to them. These antibodies may reach very high levels, and are often the first detectable indication of the disease.

Following the initial stage of the disease, and during a period that may last for years, the immune system does a good job. It quickly drives the viruses to low levels, even among the types of white blood cells that they preferentially infect. It was thought until recently that during this long period only about one in every hundred thousand of these cells in the blood is infected. New studies have shown that as many as ten per cent of susceptible cells may harbour viral chromosomes.

It was this mysterious apparent rarity of the virus throughout most of the course of the disease that has convinced a small number of biologists, notably Peter Duesberg of the University of California at Berkeley, that the AIDS virus is simply an opportunistic infection and that it has little or nothing to do with the disease itself. This suggestion has triggered an angry response from many other scientists, who point to the close association of the virus with the disease. Fewer than a hundred people have been found with an AIDS-like disease who show no signs of the viruses. Most tellingly – and tragically – many haemophiliacs who have been treated with clotting factors that were derived from large numbers of unscreened donated blood samples have developed antibodies to the virus and eventual AIDS. Duesberg continues to ignore or misinterpret these and many other facts about the virus, and it is possible that his claims have actually contributed to a recent return to high-risk activity among some American homosexuals.

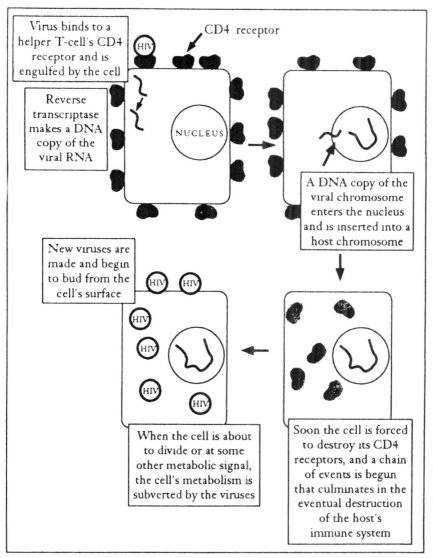

FIG. 10-1 A simplified diagram of the HIV life cycle

Destroying the immune system

Recent work from two laboratories has clearly shown that during the long and apparently quiescent phase of the disease a grim war to the death is actually being fought out between the viruses and the host's

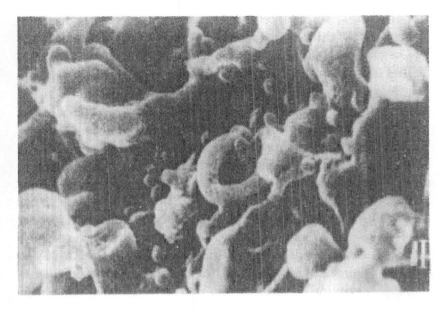

FIG. 10-2 HIV budding from the surface of an infected cell

immune system. Even though there are no obvious outward manifestations of this war, both the viruses and the cells that they infect are reproducing at their maximum rate and are being destroyed just as rapidly. Eventually, and inevitably, the viruses win, because they manage to destroy their host's cells at a rate slightly faster than the body can replace them.

Why does the immune system almost always lose this battle with HIV, when it initially seems to have the upper hand? Many other disease organisms – those that cause tuberculosis and syphilis are examples – can manage to frustrate the immune system, by sequestering themselves for long periods in parts of the body where they cannot be attacked. But HIV subverts the immune system itself, gradually altering it and destroying its function.

The AIDS viruses attack two of the three major ways in which our immune systems protect us from infection. The one that they do not attack, at least directly, is called humoral immunity, and it forms our first line of defence against bacterial and many viral infections in the blood.

As the immune system matures, the bone marrow produces white blood cells that are capable of making an astounding variety of anti-bodies. Each of these white blood cells, called B-cells, makes only one type of antibody, with which it decorates its cell surface. Then it spends the rest of its lifetime, which may be decades long, drifting around the circulatory system and waiting for a molecule to come along that its antibody will recognize. The molecule that it recognizes is called an antigen, and for reasons that will become apparent later the B-cells are most likely to recognize foreign antigens that come from outside the body. The B-cells are so diverse that some of them probably carry antibodies that would be capable of recognizing antigens from Mars.

So it is hardly surprising that most of these cells, unlike Will Rogers, never do manage to meet an antigen that they like. But if one of these cells does meet an antigen that its particular antibody happens to recognize, it is immediately stimulated to divide and to produce large numbers of identical daughter cells. A kind of genetic fine-tuning then occurs in these cells' antibody genes, which makes their antibodies an even better match to the antigen.

Louis Pasteur was catapulted to fame in 1885 after he succeeded in saving a number of children who had been bitten by rabid dogs. The inoculations worked most of the time because he was able to harness the victims' humoral immune systems. Pasteur began by growing the viruses repeatedly in the brains of rabbits, which inactivated them. Then, through a painful series of inoculations, the children were given ever-increasing numbers of the inactivated viruses. Their immune systems were led by gentle stages to make larger and larger quantities of antibodies that were specific to the viruses. The treatment worked even though the children had already been bitten – luckily, the viruses grow slowly, and Pasteur was able to train the victims' immune systems in time to slow their invasion and eventually to destroy them.*

Because the elaborate machinery of humoral immunity is left largely untouched by the AIDS virus, its victims rarely succumb to blood-borne

---

* Gerald Geison of Princeton University has recently been able to examine Pasteur's notebooks, and finds that at the outset of his work he used untried and dangerous vaccines. These may have contributed to the deaths of two patients, and his assistant Émile Roux had refused to be a party to some of the procedures. Will Pasteur's position in the pantheon of science be much disturbed? I doubt it. The method he eventually settled on has saved thousands of lives, and his process of pasteurization has probably saved millions.

bacterial infections such as those caused by staphylococci and strepto-cocci. They also remain able to make large quantities of antibodies to the virus itself. The antibodies do little good, however, once the viruses are sequestered inside the cells of the body

The second component of the immune system, and the first that the AIDS virus attacks, is also the least understood, even though it is enormously important. It is called mucosal immunity, and it provides a barrier against microbes that unavoidably enter our bodies from the outside world as we eat and breathe.

Mucosal immunity consists of a series of fall-back defences. The first is a set of rather non-specific antibodies, found in the moist membranes covered with mucus that line portions of our lungs and our guts. If this defence is breached, another type of antibody-producing cell comes into play, recognizing that there has been an invasion and stimulating specialized cells called mast cells to secrete histamine and other com-pounds. Inflammation results, and antibodies and cells capable of destroying the invader pour into the area. It is when this second barrier is breached in turn that bacteria can begin to damage our intestines or our lungs. And it is this barrier that is first breached in African victims of AIDS.

Doctors from the Liverpool School of Tropical Medicine, the Uni-versity of Oxford, and the Kenya Medical Research Institute have inves-tigated the kinds of opportunistic diseases that strike African AIDS victims. They find that the first infections are those that would normally be kept at bay by the mucosal immunity barrier. These include pneumo-coccal pneumonia and tuberculosis, in which the lung tissues are attacked. But the infections that do the most damage are a wide spec-trum of debilitating diarrhoeal diseases, which cause the victims to waste away – hence 'slim disease', a common term for AIDS in Africa.

The most widespread opportunistic infection of African AIDS vic-tims is caused by *Salmonella typhimurium*. This close relative of the terrible *Salmonella typhi*, you will recall, will usually cause disease in mice, not humans. But in African victims, within a year or two of infection with HIV, these normally harmless bacteria can begin to run rampant. Because diseases with similar symptoms are so common in African populations, the association of *S. typhimurium* with AIDS was not understood until recently.

There are a number of possible explanations for this swift and dramatic set of consequences of HIV infection in Africa. Since life-threatening enteric infections are far commoner than they are in the developed world, these are the pathogens that are most likely to be the first to breach the African AIDS victims' weakened immune systems. Alternatively, it may be that AIDS viruses interact with the mucosal immune system in ways that have yet to be understood. If so, because of the prevalence of enteric diseases, this would have a disproportionate effect on African victims. Little research is being done to explore these questions – which is a symptom of the small amount of attention paid by AIDS researchers to the seriousness and relevance of the African AIDS experience.

The other immune mechanism that is attacked by HIV, cellular immunity, deals with the most subtle and cunning set of the body's foes, and perhaps as a result it is very complicated. Cellular immunity is the province of several important classes of white blood cells, collectively called T-cells. These cells are manufactured in huge numbers in the thymus, a huge lymph-node-like gland found in the necks of babies and children which shrivels away almost completely around the time of adolescence. (Other animals have thymuses too, and it is these glands that you are eating if you dine on sweetbreads.) We go on making T-cells throughout life, however, for before the thymus disappears it has managed to seed other lymph nodes in our bodies with versatile cells that can go on dividing and producing T-cells for the rest of our lives.

There are many kinds of T-cells, but the two major kinds are killer and helper T-cells. These cells are very sensitive to changes anywhere in the body that take place as the result of an infection. When killer cells are alerted to a foreign protein on the surface of an infected cell, they can destroy that cell immediately, even if it belongs to the host. They can detect and destroy cells that have been altered only slightly, such as those that have been infected by a virus. This is the most direct kind of cellular immunity.

Helper T-cells, although they are also alerted by similar mechanisms, do not destroy the foreign cells directly, but release proteins called cytokines. These proteins in turn stimulate increases in the numbers of those B-cells that happen to make antibodies which can recognize the changes in the infected cells. These antibodies coat the infected cells, rendering them even more susceptible to the fatal attentions of the killer T-cells. Helper cells have also been shown to be instrumental in the production of killer T-cells. So helper cells, like killer cells, contribute to cellular immunity, though in a more roundabout fashion. And it is these helper T-cells that are specifically targeted by the AIDS viruses.

In a logical world, one might think, it would be the killer T-cells that the AIDS virus would attack and damage, since it is these cells that can directly detect and destroy the cells of the body that harbour the virus. If the killer cells were destroyed, this would presumably render the victim helpless to ward off viral infections, and would leave the viruses free to multiply. But it seems that this may be too draconian a move, endangering the host too much.

It may be that killer T-cells are so essential to our survival that if a virus were to attack them we would die before the virus had a chance to spread. There is some evidence that this is so. Mice can be manipulated genetically to destroy their ability to make killer T-cells, and these mice have been found to be more susceptible to infection than mice in which the ability to make helper T-cells has been similarly destroyed.

The invincibility of HIV

The intense interest of thousands of scientists worldwide in these cellular events can be directly traced to the fact that AIDS is essentially incurable. Most cases of HIV infection proceed with grim inevitability to full-blown AIDS. Yet a few of the people who acquire the virus have remained symptom-free for as long as fifteen years. Is this a true resistance? Probably not, for in at least some cases they have simply acquired a less virulent virus. One small group of such symptomless people in Australia was found to have picked up their virus through blood transfusions from a single donor. By a number of tests, this virus was less virulent than the majority of HIV strains. Indeed, there was no evidence that this weakened virus could still be passed on to other hosts.

But real resistance may have been found by Francis Plummer during the course of his studies on Nairobi prostitutes. Some of the prostitutes, in spite of their almost daily exposure to the virus, remained free of antiviral antibodies during the two years of his study. Somehow, the viruses could not establish themselves in sufficient numbers to alert their immune systems. Why these women should have this extremely effective resistance is not clear, nor is it known whether a similarly resistant subgroup might exist in the rest of the population. Plummer has found some small differences between the virus-free women and those who became infected. The differences lie in some genetic characteristics called MHC that are centrally important to the function of the immune system. But the differences, unfortunately, are not clear-cut. No single gene has yet been found that confers resistance to the virus.

Were HIV easily destroyed, there is no doubt that interest in understanding it would wane. But because it is not, it continues to drive far-ranging explorations of cell biology and immunology that might never have been carried out without the threat of AIDS. Whole bookshelves are now filled with the results.

The malaria plasmodium, with its thousands of genes and complex life cycle, is far more astounding and versatile in its effects than HIV. If a similar amount of research were devoted to plasmodia, the findings would fill entire libraries rather than mere bookshelves. But even though it has a tiny chromosome, HIV has proved to be quite sufficiently ingenious at defeating all attempts to find a cure. The more the nine genes of HIV are studied, the more ways are found in which they interact with the cells of their host.

In its early stages of infection, the virus performs subtle manipulations of events in the host cells, the effects of which then become greatly multiplied and spread beyond those cells that are actually infected. As the virus comes in contact with helper T-cells, it attaches to a molecule called CD4 on their surface. CD4 is essential to the function of these helper T-cells, for it acts as a kind of sensor that normally alerts them to the presence of something foreign.

CD4 interacts with yet other molecules, found on the surface of a variety of other cells, that form part of the MHC system. In a virus-infected cell, the MHC molecules pick up bits of viral proteins and carry them to the cell's surface. A bit of protein that is sufficiently foreign sets up an interaction between these MHC molecules and the CD4 of helper T-cells, which in turn sets in motion a series of events that results in the multiplication of cells that produce specific antibodies (Figure 10-3).

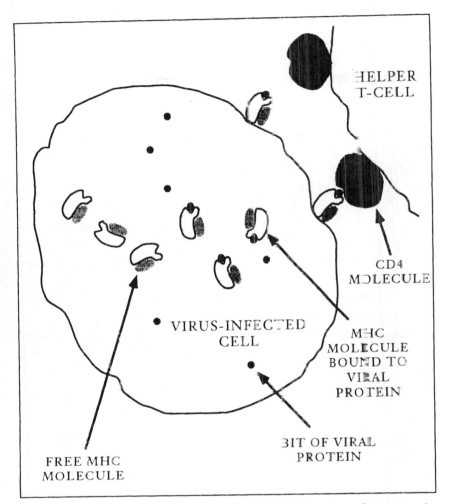

HELPER T-CELL

CD4 MOLECULE

VIRUS-INFECTED CELL

MHC MOLECULE BOUND TO VIRAL PROTEIN

BIT OF VIRAL PROTEIN

FREE MHC MOLECULE

FIG. 10-3 MHC molecules in a virus-infected cell pick up bits of virus protein and present them to the CD4 molecules on the surface of helper T-cells. This initiates a cascade of events, resulting in the production of antibodies that help to destroy the virus-infected cell. HIV interferes with this process, both by killing helper T-cells and by preventing them from making CD4 molecules.

The virus takes immediate advantage of this interaction. A large protein that is situated on the virus's surface has evolved to resemble these MHC molecules, and it is this protein that can bind to CD4 and subvert what is normally a highly specific binding process. Helper T-cells are the most tempting target, for they have the largest number of CD4 molecules on their surfaces. Macrophages also display a few CD4 molecules, and they too are attacked by HIV-1, but not as readily.

But the viruses cannot attack all our tissues. Only a few cells in the body make CD4 molecules. If any other type of cell happens to make CD4, the cell soon detects this error and destroys the molecules by engulfing them. It pulls the CD4 molecules inside itself into tiny internal pockets, within which they are digested. And it is this built-in ability, which is normally suppressed in helper T-cells, that HIV triggers when it invades.

During the early stages of HIV infection, a viral protein called *nef* forces the helper T-cells to eat and thus destroy their CD4 molecules. Later in the infection, *nef* is joined by other viral proteins that direct the cells to destroy any new CD4 molecules that they might have made even before they reach the cells' surface.

These processes, which would be quite normal in other cells, have immense effects when viruses trigger them in helper T-cells. These damaged cells now send inappropriate signals to uninfected killer T-cells (which you remember are not invaded by the virus) and stimulate them to start multiplying. The cells that result are somehow abnormal – they cannot carry out their policing function properly. It is not known whether these ineffective killer T-cells are always present in the body in low numbers and are singled out to multiply by that inappropriate signal from the infected helper T-cells, or whether the signal somehow transforms normal killer T-cells into ineffective ones. Whatever the mechanism, the influence of the AIDS viruses soon begins to reach out beyond the cells that they infect, like the spreading ripples in a pond. Over time, the keen cutting edge of the immune system becomes blunted, making it progressively less and less responsive to dangers from outside the body.

By the last stages of the disease, more highly visible changes happen. Helper T-cells, which have been gradually disappearing during the long early course of the infection, now plunge to very low levels. Even killer T-cells begin to die in unusual numbers. And during these last stages, the immune cells of the lymph nodes become highly abnormal, fusing together into great clumps. Ever-increasing numbers of opportunistic pathogens, including some that formerly were rarely seen to invade humans, can penetrate the host's rapidly shredding defences.

AIDS has revealed for the first time the true extent of the penumbra of potential infections that our cellular and mucosal immune systems keep at bay, and how absolutely essential these protective shields really are to our survival.

The HIV family tree

The AIDS epidemic has raised far more questions than we can begin to answer. Why has it happened now rather than at some time in the past? Where did it come from? Do the AIDS viruses have points of weakness that we can attack – are they, like the other epidemic diseases that we have met so far, living on the edge of what is possible for them? What are the chances that they can evolve into something even more dangerous, or are they already as dangerous as they can manage to be?

These retroviruses have not evolved in a vacuum, but are part of an extensive family of viruses that attack many different organisms and have many different effects. Even the two human immunodeficiency viruses, HIV-1 and HIV-2, are clearly and dramatically distinct from each other. HIV-2 was first found in 1985, by a group of virologists working in Senegal, West Africa. They detected it by using antibodies that had earlier been used to detect HIV-1, which means that the two viruses are close enough to provoke similar reactions from their hosts' immune systems. And yet their differences are profound

In spite of its superficial similarity to HIV-1, HIV-2 takes much longer to wreak its damage on the immune system, and the damage is at least quantitatively different – there is not as marked a decline in the number of helper T-cells as there is during an HIV-1 infection. A five-year study comparing the effects on two groups of women, each infected with one of the two viruses, was recently carried out in West Africa. By the end of the study one-third of the group infected with HIV-1 had died of its effects and many others were in advanced stages of the disease, while all the members of the group infected with HIV-2 had survived and their symptoms were much milder.

The effects of the HIV-2 virus may take decades to develop, and so far it does not seem to pose the immense threat to the human population that HIV-1 does. Further, it spreads from one person to another much

more slowly than HIV-1. While it is common in West African populations, it has not spread to the rest of the world with the same rapidity
as HIV-1. Cases of HIV-2 infection in the US and Europe are very
rare.

None the less, HIV-2 is by no means harmless. It has recently been
shown to produce an AIDS-like disease in baboons. The course of
the disease is very swift in these animals, with the first appearance of
opportunistic infections within eighteen months. Some of the baboons
used were of a species that is common in West Africa, and it is surprising
that they are so seriously affected by a virus that is also common among
the people in the region.

When their RNA sequences are examined, the two AIDS viruses
turn out to be very different. You cannot look at the sequence of a gene
from HIV-1 and mistake it for a gene from HIV-2. On average, the
RNA sequences of the two viruses are only sixty per cent identical, and
in some genes, like *nef* and the gene that codes for the viruses' envelope
proteins, the identity drops to a mere forty per cent.

This low identity makes it very difficult to determine where the
viruses came from, and when they diverged from their common ancestor. Luckily, the question of their evolutionary origin has been clarified
by the discovery of yet another, closely related, class of viruses, this
time in monkeys. In the same year as the HIV-2 virus was discovered,
a group of scientists at Harvard reported the isolation of a virus that
was causing immunosuppressive disease in a laboratory colony of Asian
macaques.

Similar viruses, named SIV for Simian Immunodeficiency Virus, have
since been isolated from many different wild populations of African
monkeys and apes. These SIV viruses are so common that it seems
certain that they do little or no harm. They are particularly prevalent
among green monkeys.* Over half of the monkeys in some parts of
Africa can be infected. Other SIV strains have also been found in some
of our closer relatives, including the chimpanzees, although the last are
very similar to HIV-1. It has been suggested that they may be infections
picked up in the past by the chimpanzees from humans, but if this is
so, as we will see, it must have happened some time ago.

---

* 'Green monkey' is actually a common name for a number of species of long-tailed
monkeys that are found throughout sub-Saharan Africa.

The various SIV strains are found in wild populations of African monkeys, but they have not been found in wild populations of Asian monkeys. It seems that the Asian macaques being studied by the Harvard scientists had picked their viruses up from African monkeys in the same primate colony. Further, the African monkeys and apes show no ill effects from SIV infection, but Asian monkeys that have been deliberately or accidentally infected with SIV rapidly get sick with an AIDS-like set of symptoms. All this suggests that the simian viruses are most likely to have originated in Africa.

So how are all these types of virus related? Confronted with the evidence of the molecules, evolutionists would conclude that the HIV-2 from Asian monkeys and some of the SIV strains, since they resemble each other so closely, must have had a very recent common ancestor. But HIV-1, the disease that is causing the great AIDS epidemic, appears to have evolved for a slightly longer time on its own separate path. I have used long stretches of their RNAs to construct an evolutionary tree, shown in Figure 10-4, that demonstrates these relationships quite clearly.

I made the tree using sequences of the reverse transcriptase gene and the *gag* gene, which are among the most slowly evolving of the HIV and SIV viruses' genes. And I used a few tricks of my own to collapse a large number of sequences into a more manageable number, and at the same time to get an accurate idea of the lengths of the various branches. The longer these branches, the greater the amount of evolution that has separated them.

The first thing that you will notice is that HIV-1 and HIV-2 are very far apart on the tree. Lying between the two groups of human HIVs are two major branches of SIVs, one group coming from African green monkeys and the other from a distantly related Sykes' monkey from East Africa and a mandrill baboon. All these SIVs have diverged so much in the course of their evolution that they bear only a remote resemblance to each other and to the HIV-1 and HIV-2 sequences. But quite near HIV-2 on the tree is a cluster of SIVs from Asian macaques and African sooty mangabeys. Some of the macaque SIVs are essentially indistinguishable from SIVs from the mangabeys, strong evidence that they are the result of a recent accidental introduction – in this case from the mangabeys and other monkeys with which they shared their captive quarters.

The chimpanzee SIV shown in the tree is some distance away from both of the two major types of HIV-1 (and it is striking that the chimp carrying this virus has yet to suffer ill effects from it). If this virus really

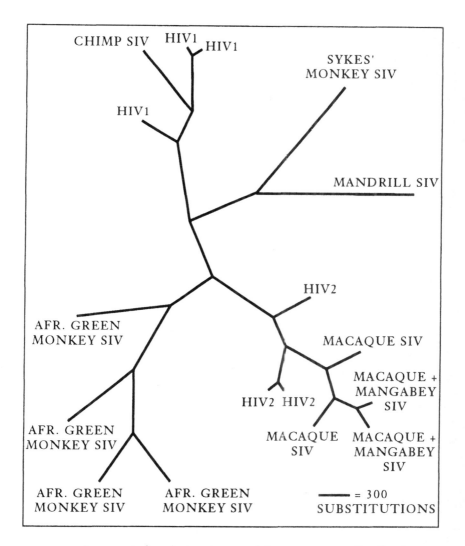

FIG. 10-4 An unrooted evolutionary tree of the sequences coding for the reverse transcriptase and gag genes of the HIV and SIV viruses. This is the shortest tree that connects all these sequences. The further apart two sequences are positioned on the tree, the less closely related they are to each other.

did invade chimpanzees from humans, it must have done so a relatively long time ago.

The diverse simian viruses, it would seem, are living a fairly well-adjusted life among a wide range of African monkeys. The great distance between HIV-1 and HIV-2 shows that they probably arose, independently, from different members of this highly heterogeneous collection of SIVs. At some point in the past, the ancestors of HIV-1 and HIV-2 made separate jumps to our own species, evolving dramatically in the process. The two very different types of HIV-1 may also have infected our species independently, and in the case of HIV-2 there may have been a separate jump into the mangabeys.

It is interesting that the HIV-1 branch is more than twice as long as the HIV-2 branch. It appears to have diverged at least as much from its ancestor as the various SIV strains have diverged from each other. At first sight this suggests that it arose from its SIV ancestor a long time ago. Yet the first recorded cases of HIV-1 infection date from 1959, one from a blood sample collected in Zaire and the other from a tissue sample of a young man from north-west England who died of the disease. The young man is only known to have ventured out of England once, as far as Gibraltar, and it is possible that he picked up the virus there or in nearby Morocco.*

But it is of course true that the HIV-1 virus has been impossible to detect until recently. Because it is so clearly separated from the other retroviruses, it seems likely that it was actually infecting small numbers of humans in Africa long before 1959. It would have existed at low levels, making up a minor and unnoticeable part of the plethora of diseases in that sad and luckless continent. Then came societal upheaval on a grand scale, and a new opportunity for the virus.

---

* A recent attempt to reisolate the virus, which is suspiciously modern in its sequence, failed, so this particular instance of an 'early' HIV is in doubt. But the Zaire sample may still be from the 1950s, though unfortunately sophisticated checks cannot be made because the sample has been used up.

Evolution at warp speed

The accuracy of the evolutionary tree, of course, depends on the assumption that the molecules are telling us a true story. The evolution of these retroviruses is remarkably swift, happening a million or more times faster than the rate of evolution in most animal and plant genes. It usually takes tens or hundreds of millions of years for ordinary genes such as the ones inhabiting our own chromosomes to accumulate as many differences as those that are found among the genes of HIV-1, HIV-2 and SIV. Compared with the genes on our chromosomes, all these retroviruses seem to be in a kind of evolutionary overdrive.

This is because the enzymes that they use to make copies of their chromosomes are extremely clumsy, often putting the wrong bases into the growing RNA chain. To compound the problem, the enzymes then ignore their mistakes and blithely continue with their copying as if nothing were amiss. In contrast, the DNA polymerases that our own cells use to manufacture copies of our chromosomes are far more accurate. While our DNA polymerases occasionally do insert the wrong base, they have the capability to proofread the sequence they have just constructed and then correct the mistake.

The far greater ineptitude of the viral enzymes causes mutational mistakes to accumulate rapidly. They can reach substantial numbers within a few months of the invasion of a host. A monkey that has been infected with a single genetically uniform strain of SIV rapidly accumulates a bewildering variety of different strains. The chemist Manfred Eigen has called this great crowd of types a 'quasispecies'.

The most important result of this genetic lability is that it is almost impossible to construct a vaccine against the viruses. Further, it has also proved to be extraordinarily difficult to determine which of these innumerable changes actually matter to the viruses and which are simply evolutionary noise. The problem is equivalent to picking out someone who is singing 'Take Me Out to the Ballgame' from a noisy crowd in Dodger Stadium. If we had some idea of the proportion of all these changes that are actually meaningful, perhaps we could pinpoint the parts of the viruses that interact most directly with our immune system and that allow them to penetrate our defences.

I recently took a stab at this question myself. I used a statistical test, developed by the population geneticists Takashi Gojobori, Austin Hughes and Masatoshi Nei, to find out which of the changes in the viral chromosomes are meaningful and which are not. The test depends on the fact that if one base in a gene changes to another through mutation, the result is not always a change in the protein that the gene codes for. Most of the time there is a change, but sometimes there is not.

This is because some amino acids are coded for by more than one codon. If GGG is changed to GGA or GGU or GGC, none of these changes will have any effect on the protein, since all four codons specify the amino acid glycine. But if GGG changes to GAG, then this will be a meaningful change, resulting in the substitution of a glutamic acid for the glycine at that place in the protein.

Meaningful and non-meaningful changes are happening all the time. Hughes and Nei pointed out that if meaningful base changes are found to have happened more often than would be expected by chance, the protein itself must be evolving rapidly in a meaningful way – that is, selection must be strongly pushing the protein in a particular direction. But if they happen less often than would be expected by chance, then selection must be playing a conservative role. This conservative kind of selection weeds out many of the meaningful changes because they are harmful. The majority of the changes that survive this weeding process are those that have no effect on the composition of the protein.

I found that in some cases most of the differences between two viral sequences are meaningful changes. The most striking effect is seen in one class of mutations called transversions, and the most dramatic effect of all is seen in the *nef* gene of HIV-1 – the same gene, you will remember, that persuades a cell infected by the virus to eat its own CD4 receptor molecules. Virtually all of the transversion changes that separate closely related *nef* genes are meaningful. The effect is almost as striking in the *env* gene of the same virus, which makes the protein on its surface that interacts with the CD4 receptors on helper T-cells. In short, these genes are being pushed by natural selection to evolve like crazy.

As these viral genes continue to evolve and become more separated from each other, however, the innate conservatism of the evolutionary process reasserts itself, and the proportion of meaningful changes drops.

What does this mean? Apparently, the first response of the viruses is a burst of change, particularly in the genes that are involved in interaction with the host's immune defences. As a result, with each new infection, a great crowd of viral quasispecies is being generated at an enormous rate. This suggests that it must be very difficult for these viruses to overcome their hosts' defences, so that they must call up all their resources in order to do so. Indeed, so rapid is this pell-mell change that many of the viruses isolated from patients with full-blown AIDS are defective, and are unable to escape to a new host.

The chance of an infection being passed on is a thousand times greater per sexual contact during the first few weeks of the infection than it is during the long latent period when the virus is battling the host's immune system. This suggests that the most frantically evolving viruses, caught up in the battle with the host, are not usually the ones that go on to infect somebody else. Instead, the viruses that are passed on are likely to be those that have not yet undergone this evolutionary burst.

I suspect that, just as with so many of the other pathogens we have looked at, there is a trade-off operating with HIV-1. As the ever-diverging strains of this virus battle the immune system of their current host, most of them lose the ability to infect new hosts – they are simply not resourceful enough to carry out this battle and still retain full infectivity.

This means that the destruction of the host's immune system may be an accident, possibly a by-product of the fact that the viruses have multiplied to such high levels at the outset of the infection. Once a sufficiently large number of viruses is introduced, and a sufficiently large cloud of quasispecies develops, viruses and host cells join in a battle that has a grimly inevitable outcome. The outcome is fatal to the viruses as well, for the viruses that finally win this conflict will perish with the deaths of their hosts.

Why so many viruses at the outset? Presumably for the same reason that the bloodstreams of victims of bubonic plague or falciparum malaria swarm with so many pathogens. Sheer numbers are required if some of them are to make the transfer to new hosts. Similarly, sheer numbers seem to be required if a person infected with HIV is to pass on the virus. The strains that are successfully transmitted are the ones that can make very many viruses quickly. It is only later that this great crowd of viruses, by mechanisms not understood, sets in motion the events that bring about the eventual destruction of the host's helper T-cells. This destruction probably has little to do with the survival of the viruses as a species, for it usually takes place after their ancestors have already managed to spread to new hosts. Syphilis, too, is most infectious in the early stages of the disease, and most of the damage caused by the syphilis spirochete happens long after this early infectious period.

All of these retroviruses, particularly HIV-1, seem to be working at the very limit of what is possible for them. Each new infection of a human or monkey host challenges their genetic resources to the utmost. Because of this, I do not think they are likely to take up some dramatically new mode of transmission, the prospect of which so worries epidemiologists. If they were to do so, they would probably have to change so much in the process that they would constitute far less of a threat to our immune systems.

Battling HIV

What can be done to fight this most insidious of enemies? Antiviral drugs such as AZT, which interfere with the replication of the viral RNA, have only a small effect. They are even dangerous, for they interfere with normal RNA synthesis in the cell as well. Antibodies have proved ineffective, for two reasons. First, the viruses spend most of their time hiding where antibodies cannot reach them, and the lack of helper T-cells in the host means that in any case they cannot be detected in those hiding places. Second, they evolve with such blinding speed that they are soon making very different proteins from those that were made by their ancestors a few weeks or months earlier. They rapidly become invisible to any antibodies capable of interacting with the proteins of those ancestors.

Highly specific drugs, for example modified RNA molecules that can actually split the viral RNA into pieces, are in development, but while these are effective in cells in a laboratory dish it is unclear how well they will work in the human body. Various tricks are currently being used to target these drugs to the tiny minority of infected cells.

All these efforts are stopgaps, hindered by the fact that HIV is far more difficult to kill than bacteria. Once HIV incorporates itself into our chromosomes, it is like trying to kill a piece of ourselves. The real scientific solution will come from our rapidly-growing understanding of the immune system and how it works. It will be essential, for example, to understand the nature of the mysterious signals that the tiny minority of infected cells send out to the killer T-cells, the signals that pervert and eventually destroy the capacity of these cells to fight off cellular infections. If those signals could be blocked, it might not matter if the viruses infect a few cells.

However, even if a high-tech cure becomes available, it is unlikely to penetrate very quickly to millions of people in the Third World who need it. This means that becoming infected with the HIV-1 virus will remain an irreversible and almost certainly fatal event for most of its victims in the near future. So at the moment, the most effective measures to halt or at least slow the epidemic are those that prevent an infection from happening in the first place.

As I travelled around India, where the AIDS epidemic is in its early stages, I was told repeatedly that the disease is unlikely to become a large problem there because of the puritanical structure of Indian society. Yet in southern India it is starting to spread in exactly the same way as it has done in equatorial Africa. Everywhere along the dusty roads there are small primitive truck stops, some consisting of little more than a shack where refreshments are sold and a row of aluminium cots that have been set out in the open air. Groups of prostitutes often await the truck drivers who stop, providing the same environment of casual sex that has given HIV its opportunity in Africa. AIDS is spreading rapidly within these groups. In the Indian population at large, about seven out of every thousand people are already seropositive, and the incidence is much higher among these at-risk groups.

But the real explosion of AIDS in India is taking place in the far north-east, particularly where the state of Manipur presses up against the Burmese border, the same area in which my uncle fought the Japanese half a century ago. Here the flow of drugs is so great that their use has permeated all levels of society. The problem, microbiologist Geeta Mehta told me, is compounded by the social structure of Manipur, which is extremely matriarchal. Many men are underemployed, and the use of alcohol and drugs among these idle men is rampant. In Manipur state, just as in America, dirty needles are spreading the virus. The resemblance to the situation in American ghettos is striking.

When the story of the AIDS epidemic is considered decades from now in the light of history, blame will surely be apportioned. Much of it will belong to those who tried to pretend that the problem did not exist or was irrelevant to their own segment of society, and who obstructed educational and public health measures that could with relative ease and low expense have slowed or even prevented its spread. And much more blame, as we saw earlier, will be laid at the doors of those who encouraged the societal destruction and upheaval, particularly in Africa, that first gave this rare virus its opportunity to become common.

Yet research into this epidemic may have far-reaching rewards. Among other things, HIV can teach us lessons about our own physiological response to disease. Many different diseases, ranging from the exotic like *verruga peruiana* to the commonplace like viral pneumonia, penetrate our defences by interfering with various parts of our immune systems. These and many other disease organisms can take advantage of any temporary breach in our defences that result from stress or poor nutrition. Only HIV causes irreversible damage, but in order to do so it seems to be working at the limits of its capabilities. A sign of this is the frantic evolution that the AIDS viruses begin as soon as they enter new hosts. Each victim comes heartbreakingly close to winning the battle against the virus – it is impossible to suppose that new ways will not be found soon to augment our defences and tip the balance in the other direction.

The future of our plagues

Predicting things is hard, especially about the future.
Attributed to Yogi Berra

In Maharashtra state, within days of the report of the plague outbreak, DDT was rained from the skies by helicopter, killing the occasional rat flea and a good many other creatures besides. In the coastal cities of Peru that were stricken by cholera, medical teams moved swiftly to give the victims rehydration therapy, and as a result the mortality was held to one case out of every hundred. Even in the ghastly refugee camps of Rwanda, cholera and typhus were quickly controlled by brave volunteers.

Worldwide, immunization programmes have been fabulously successful in lowering the incidence of polio, diphtheria, tetanus, mumps, measles and other diseases. Most of the ancient scourges of our species, it seems, can be controlled by the application of technology. In the most dazzling success of all, smallpox seems to have been expunged utterly from the planet.

Yet many diseases remain. We have examined some of them and found the reasons why they persist, though we have also discovered that what can be done about them is not obvious. All these diseases are in principle as vulnerable as smallpox and plague – all of them, even in the tropics, seem to be working at the very limits of their capabilities. Yet all of them continue to exact a yearly toll of death and disability that sometimes runs into the millions.

Many of these less tractable diseases are on the increase. AIDS has no cure, and threatens to become the worst epidemic in recorded history. Cases of tuberculosis are becoming more common, in part because the cure is so long and complicated that it is often never completed, with the result that drug-resistant bacilli multiply. And as we have seen, cholera is always ready to leap forth when the conditions are right for it.

Paul Ewald has made a strong point about the evolution of virulence and the future of the many diseases that are spread through the water supply. There is a huge pay-off in public health, he observes, when water supplies are cleaned up. Vicious strains of diarrhoeal diseases are replaced by milder ones, so that even when there is an outbreak the symptoms will be less severe. This is, he suggests, an evolutionary process, and it is perfectly understandable in these terms. Under conditions where transmission is easy and there are plenty of human hosts, as in the delta of the Ganges, severe strains that spread quickly can survive – they will be transmitted even though they soon kill their hosts. When the water supplies are improved, however, and transmission is no longer so easy, strains are selected for that do not kill their hosts immediately, or even do not kill them at all. In order to survive, these strains must be able to persist for long periods in their hosts, moving about with them, and producing smaller numbers of bacteria over far longer periods of time. This increases the likelihood that they can take advantage of some brief window of opportunity, some moment of uncleanliness, and spread.

All this is true enough, and it makes a powerful argument for public health measures. But it is not, in most cases, a real evolutionary process. Robert Selander of Pennsylvania State University and other evolutionary detectives have shown that many virulent strains of bacteria actually have long evolutionary histories, so that they must have persisted at low levels even when they were not being selected for. In the short term, vicious strains do not evolve into milder ones. Instead they drop in frequency, and the milder ones that were always present increase. And the process is quickly reversible, as we have seen repeatedly over the last few years. If water supplies become filthy again, the vicious strains will rapidly multiply, or even be reconstituted from the various plasmids and other component parts that are continually being passed around among the bacterial populations. And sometimes, puzzlingly, changes that have little to do with virulence occur even as water supplies are being cleaned up. It is not at all obvious why cholera strain O139, which is apparently at least as vicious as the O1 strain that it is replacing, is spreading through eastern India and Bangladesh.

This means that, as the world's population continues to grow, we will not be able to drop our guard for a moment. My guess about the immediate future is that as we move into the inconceivably crowded twenty-first century we will enter a period in which any breakdown in public health will have enormous consequences, particularly in the tropics. And there will be two kinds of outcome.

The first will happen if we learn from our experience. India and Peru have learned a severe lesson from their recent bouts with disease. Plagues can be both expensive and embarrassing. Trade and tourism can plummet. To be crass about it, having a plague does not pay. Both countries are taking some steps that will decrease the likelihood of such outbreaks in the future, though both are hampered by a lack of funds and by the immobility of entrenched bureaucracies.

The second outcome is grimmer. Many countries, particularly in sub-Saharan Africa, do not have appreciable amounts of either commerce or tourism. Some, like Somalia, lack a government as well. Intervention from the outside, although it has often had striking short-term effects in relieving disease and famine, has been difficult and dangerous. Such aid is likely to decline in the future. It is here that the epidemiological horror stories of the next century are likely to be played out. Will much of Africa plunge into a long night of disease, famine and hopelessness as profound as that which enfolded Europe during the Dark Ages? What will be the response of the rest of the world to these events, which will certainly bear out the worst predictions of Malthus? There will have to be a response – the world is getting smaller, and these days the road is too narrow to allow the rest of us to cross over to the other side.

At the same time, much of the world will continue to be plagued by the diseases that, like AIDS and tuberculosis, are resistant to cure. Some, like AIDS, appear to contradict the expectation that if transmission is difficult, severity will decrease. AIDS is an almost invariably fatal disease, but much of its transmission takes place during the early and relatively symptom-free stages of infection. Its severity appears to be a kind of accident, perhaps simply an unintended consequence of the large number of viruses that are required for successful transmission. But this is not always the case, even among the close relatives of HIV. A large proportion of African monkeys in wild populations are infected with SIVs, so these are presumably easily transmitted, yet the monkeys suffer no noticeable ill effects. This is a puzzle – why are not AIDS-like diseases common among these monkey populations?

Perhaps they are, or have been, and we simply don't know enough about these wild populations to detect them. The strain of SIV that was accidentally introduced into laboratory colonies of Asian macaques quickly spread and caused an AIDS-like disease. Such laboratory colonies, with their crowding and disturbance of normal behaviour patterns, are ideal places for new strains of virus to thrive. My guess is that, sooner or later, a mutant SIV with enhanced virulence will arise and spread in a laboratory colony of African monkeys as well. These viruses are only a step or two away from such increased virulence. You will remember that human HIV-2, closely related to the SIVs of African mangabeys, can produce an AIDS-like disease in baboons. As with so many other pathogens, new virulent strains of these viruses will arise whenever their hosts are sufficiently crowded together and there is sufficient ecological disturbance.

Falciparum malaria, too, contradicts the expectation. The parasites are transmitted through mosquitoes only with difficulty, but the disease has not become less severe as a result. Instead its severity is high, since only the strains that produce the largest number of parasites in the bloodstream are likely to be transmitted to new human hosts.

Thus it is at the point of transmission that these apparently intractable diseases have their greatest vulnerability. If transmission can be made even slightly more difficult, I predict that they will disappear or fall to low levels more rapidly than we might expect. Even small changes in patterns of sexual behaviour can rapidly reduce the incidence of new HIV infections. Small changes in living habits have had a large effect on falciparum malaria, and my guess is that Manuel Patarroyo's malaria vaccine will have an impact far out of proportion to its apparently poor effectiveness in early clinical trials.

There is no doubt that advances in molecular biology should be able to accelerate the development of antibiotics and vaccines, producing new types that will be far more effective. But it is unclear whether the large investments that are required will be made until our feet are held to the fire. For example, it is astonishing that so little can currently be done against tuberculosis.

Most cases of tuberculosis are essentially invisible, and only a small proportion move to the life-threatening and highly infectious stage. This small proportion, however, accounts for two and a half to three million deaths a year worldwide, making TB the greatest cause of death that can be traced to a single pathogenic organism.

Currently, seventy per cent of the world's children are given an attenuated live-bacillus vaccine called BCG. The vaccine was developed in the 1920s by Léon Calmette and Alphonse Guérin, who grew the tubercle bacilli for hundreds of generations in an unfavourable medium that contained bile. They ended up with a strain that has remained harmless to this day. Unfortunately, the vaccine produced using this strain works rather poorly. Some studies have shown a seventy per cent effectiveness rate, others no effect at all, and a good many have fallen somewhere in between. Still, the vaccine turns out to have most surprising properties. A recent study carried out in Malawi in central Africa found no effect of the vaccine against tuberculosis, but an almost fifty per cent effectiveness against the closely allied disease of leprosy.

In First World countries, where the incidence of TB is low, use of this vaccine is rare because its effectiveness is so uncertain, and because after it has been given, the skin test for the presence of tuberculosis no longer works. This means that once somebody has been immunized using BCG, not only is there no guarantee that he or she will be protected, but there is no easy way to determine if a later infection with real tuberculosis bacilli might have taken place. Further, BCG cannot be given to AIDS patients, for their immune systems have been destroyed to the point that even this attenuated bacillus can multiply in their bodies, giving rise to systemic infection. This doubtful immunization, along with a complex armamentarium of drugs that must be administered over a period of months, are the only defences that we currently have. This technology, which has not materially changed for decades, cries out for improvement.

Clever thinking is demanded. Mucosal immunity, that least understood part of the immune system, is central to our ability to fight off the invasion of the tubercle bacillus through the lungs. It has recently been found that giving BCG in aerosol form to guinea pigs, so that they breathe it in the form of a mist, stimulates their mucosal immune systems. This greatly increases their ability to fight off subsequent infections of the unmodified and dangerous tubercle bacilli. One wonders whether BCG might have been administered in the wrong way to all those hundreds of millions of people during the sixty years since it was invented – should we all be given aerosols of the bacteria to breathe?

The challenge of TB raises the question of how useful our immunological and biochemical defences will continue to be. For example, will antibiotics lose their effectiveness as resistant strains of bacteria are selected for and replace the susceptible strains? The prognosis is mixed. In India, a wide variety of antibiotics are available over the counter, as they are in many other countries. Researchers at the Lady Hardinge Medical College and elsewhere have been monitoring changes in drug resistance over time. Although resistance to antibiotics among *Shigella* and *Salmonella* strains has risen rapidly in recent years, it is now beginning to decline as doctors prescribe the antibiotics more cautiously and some regulation has been imposed on their sale. But antibiotic-resistant strains of the tuberculosis bacilli have continued to increase in numbers, primarily because it is so difficult for many patients to complete the course of therapy.

The problem of pathogen resistance in TB is severe and is likely to continue for a long time. In the US, strains of the TB bacillus that are resistant to at least one drug make up about ten per cent of the total, while in India an astonishing seventy per cent of the strains that have been isolated have proven to be resistant to the drug isoniazid. And in both countries, a majority of AIDS patients develop active TB – over eighty per cent of them do so in India, where TB is as prevalent in the entire population as it is among the homeless of San Francisco.

There are some rays of hope in the tuberculosis story. These bacilli, unlike so many of the bacteria that cause diarrhoeal disease, do not acquire their drug resistance from plasmids and gene clusters, for they are not able to pick these up from other bacteria. All the drug resistant strains examined so far have mutations that inactivate specific enzymes or that damage part of the cellular machinery. In the absence of anti-biotics, these resistant strains are likely to grow rather poorly. A good indication that they do not survive for long is that hundreds of different drug-resistant strains have been isolated, each one apparently the result of a different set of mutations. If any one of these were especially long-lived, it would tend after a while to dominate the population of drug-resistant bacteria and perhaps even take over the whole popu-lation. Luckily, this seems not to be happening.

There is no doubt that effective defences against even drug-resistant pathogens can continue to be generated, since our knowledge of these pathogens will continue to grow. But as these drugs become more sophisticated, they are also becoming progressively more expensive. They are being priced out of the reach of the Third World, as has already happened with mefloquine and the more exotic antibacterials. Prevention, through public health measures, education and changes in living habits, may soon be the only practical route the Third World can take.

It is striking that many diseases spread most easily in rural conditions. People in tropical rural areas are often bitten by a great variety of malaria- and dengue fever-carrying mosquitoes, and by flies that carry diseases such as leishmaniasis, Japanese encephalitis and kala-azar. They bathe and work in bodies of water inhabited by snails that carry the widespread and debilitating parasitic disease schistosomasis. Migration to the cities has reduced the incidence of these diseases in many parts of the world, but at the same time this crowding may increase the chance of an outbreak of diarrhoeal disease.

Of course, the First World public is most concerned with the possi-bility of the sudden appearance of a previously unknown disease that sweeps the planet and kills millions in a latter-day reprise of the Black Death. This is I think very unlikely though it makes for good entertainment. A recent film, *Outbreak* starring Dustin Hoffman, received mixed reviews. Its premise was that a viral disease brought into the US by an African monkey spreads like wildfire through the human population, only to be controlled at the last minute. Unfortunately, the film was flawed by a spectacular piece of miscasting – the part of the African monkey was played by a cute little *Cebus* monkey from South America!

Luckily, most of the emergent viruses that have been tracked over the last few decades either cannot spread from one human to another or spread with the greatest difficulty. Our species may not always be so lucky, but there is no doubt that if a new virus emerged that killed rapidly and spread easily, it would soon be halted by the frenzied marshalling of the entire armamentarium of modern epidemiology and medicine – as happened with India's outbreak of plague. There is of course nothing to prevent the emergence of new diseases that like AIDS are slow to develop and initially invisible. But I have argued in this book that even such diseases are surprisingly vulnerable if we keep our wits about us and attack them where and when they exhibit their greatest weakness.

Many people, biologists among them, have suggested that our species is headed for a drastic Malthusian solution to the growing problems of population growth and ecological damage. We are so crowded, and have damaged our environment so much, that surely some new plague or plagues will sweep through the planet, decimating our species. It will fall to those of us who survive to rebuild the shattered remnants of civilization and (perhaps) learn the perils of reproducing too quickly and damaging the environment too much. Such a dramatic reduction in population, accompanied by great societal changes, took place in Europe during the Plague of Justinian and later during the Black Death. Why shouldn't a similar disaster happen today?

I am confident that no terrible disease will appear that slaughters us by the billion. The reason is that we can now respond very quickly to such a visible enemy. Any disease that spreads like wildfire will have to do so through the air or the water, and there are many steps we can take right away to prevent such a spread. If the people of fourteenth-century Europe had known what we know now, they could have halted the Black Death in short order.

The most insidious and dangerous diseases at the moment, and the ones about which we can do very little, are the largely invisible ones such as AIDS and tuberculosis, and the all-pervasive ones like the many-headed hydra of diarrhoeal illnesses that thrive on poverty, ignorance and war. Yet these diseases, disastrous as they are, will have surprisingly little effect on the immediate future of our species. The World Health Organization has estimated that deaths from TB are likely to rise from 2.5 million per year at the present time to 3.5 million per year by the end of the century. It is entirely possible that the AIDS epidemic will eventually account for a similar number of deaths. But immense as these numbers are, and disastrous as these diseases have proven to be, their effects are swamped by population growth. The world's population is currently increasing by at least a hundred and twenty million a year, a number greater than the entire population of Europe during the Middle Ages.

We will not be handed a dramatic solution to the problem of over-population by a great plague that wipes most of us out. We will have to solve it through the unglamorous but essential processes of societal change and education.

But at the same time, while we can undoubtedly control our own plagues, at least in the short term, we will soon be confronted by the accumulating impact of plagues in other species. No species is an island.

11

Safety in diversity

Motley's the only wear.

Jaques, in Shakespeare's *As You Like It*.

Consider the AIDS viruses once again, and their fierce struggle against the immune system of each person they infect. Much of that viral evolution appears to be a response to the fresh challenge that each victim poses. Host diversity is a barrier to the spread of the disease, a barrier that can only be overcome with great difficulty. Indeed, some people, to judge from Francis Plummer's discovery of the antibody-negative Kenyan prostitutes, seem to be completely resistant to attack by the virus. Why should there be so much diversity among us, and what does this tell us about the way that we have evolved along with our diseases?

In addressing these questions we will confront the most astonishing story of all. So far in this book we have dealt with day-to-day skirmishes between ourselves and our diseases. It is now time to back away and look at the whole battlefield. We will discover an endless war that has shaped not just the diversity of our own species but that of the entire living world.

A few years ago, the immunologist Douglas Green and I began to wonder about the evolutionary consequences of diseases. Our thinking led us through a maze of possibilities, and down quite a few garden paths. But gradually a fascinating pattern began to emerge. Disease resistance and susceptibility in our species is incredibly complicated, and many different genes are involved. None of us, luckily, is susceptible to all the diseases that surround us, but each one of them is capable of claiming some victims among us.

Gradually Doug and I began to home in on the interesting questions. Is all our mind-numbing genetic complexity a sign of important evolutionary events in our past? Are we somehow its beneficiaries? Do our genes, and the different genes of our neighbours, somehow manage to protect us against the penumbra of diseases that surrounds us?

There is a phenomenon called *herd-immunity*, which is well known among farmers and ranchers. They have found that it is not necessary to immunize all the animals in a herd against a disease – if a majority is immunized, the disease is as likely to disappear as if the entire herd had been protected by immunization. This is because the overwhelming preponderance of immunized animals protects the remainder, driving the disease organisms down to such low levels that they can no longer spread from one of the thinly-scattered susceptible animals to the next. The resistant majority protects the still susceptible minority.

Are we, like susceptible members of a herd of domestic animals, somehow protected by the resistant people who share our lives? Is our entire species, in short, the beneficiary of a kind of herd-immunity, a herd-immunity that is built into our genes?

Formulating this question was a long and complicated process. At first we were confused and rather intimidated by the complex way that our genes and the environment interact to produce resistance or susceptibility to each of our plethora of diseases. Diphtheria, for example, is spread easily from one child to another in crowded households, but adults who have been exposed to the disease at some point in their lives, or who have been immunized against it acquire very effective lifelong immunity. In order for the disease to spread, there must be a susceptible subpopulation that is numerous enough to overcome this herd-immunity and make the transmission of the disease likely. If that subpopulation is large enough, and the bacteria that cause diphtheria are plentiful enough, then the disease is able to spread even in the face of widespread immunity. This is what is happening in the countries of the former Soviet Union today.

If there are many unimmunized children in a household or in a community, they make up the subpopulation that is at substantial risk. But not all the unimmunized children are equally at risk. The bacterium has another hurdle to overcome, for it can only infect the susceptible children, a subpopulation of the subpopulation.

Why are some children susceptible and others not? The answer lies in genetics. The human species carries a huge pool of genetic variation, and, as Philip Murphy showed, a surprising amount of this variation is found among the genes that protect us from disease. Attempts have been made, by Robert May, Roy Anderson and others, to come to grips with this variation and what it might mean for disease resistance and susceptibility. As they discovered, and as Doug and I soon found, attempts to model this complexity can quickly become very difficult.

We began with what we thought might be some simple assumptions. We used a computer to construct a simulation of a host population and supplied it with a series of *alleles*, different forms of a gene, each of which conferred resistance to one out of a set of diseases. No member of the host population could have more than two of these alleles, one inherited from the host's father and the other from the host's mother. But the population as a whole could of course have more than two alleles. Over time, due to the vagaries of chance, some of these alleles will be lost from the population. Could we come up with a simple model that would slow this loss and maintain most of these alleles over long periods of time, even though the pathogens that were causing these diseases were themselves continually evolving? If genetic variation in the population for susceptibility and resistance really was advantageous, then there should be selective pressure to maintain it.

The assumptions underlying this model seemed reasonable to us, because we knew of a great variety of different kinds of genetic variation in the human population that are like the genes we modelled in the computer. Some of these genes are those that cause us to have different blood groups. One particularly important blood group gene has three common forms or alleles in the population, named A, B and O. But none of us can have more than two of these alleles, and many of us have only one. The ABO gene, as we will see shortly, is one of many that are known to have a role in disease resistance and susceptibility, but this role is only now beginning to emerge after almost a century of study. It turns out that I am blood type O, which means that I am *homozygous*, with two copies of the O allele, while my daughter has blood type B and is *heterozygous* – she has inherited an O allele from me and a B allele from my wife. My wife is blood type B, but she might be homozygous for B or heterozygous for O and B, since the B allele is dominant over the O allele. None of us carries allele A, which is dominant over B.

And none of us, of course, is able to have an O, an A and a B allele simultaneously. Although nobody in our immediate family happens to have the A allele, plenty of people more remotely related to us do. Genes like ABO, which exhibit a number of alleles, are called *polymorphic* genes – they come in many forms. Other genes that do not exhibit all this variation in the population are known as *monomorphic* genes – if they do have any variation at all, this is usually in the form of rare mutant alleles that affect only a few individuals in the population.

Doug and I knew that many of the genes that are directly connected with our bodies' defences against disease can indeed be astonishingly polymorphic, far more so than the ABO blood group genes. Some of these genes are much better understood than others. One particularly striking set of these polymorphic genes is clustered together on our sixth chromosome. Together they form a multipurpose group of genes that we met briefly in the last chapter, named the major histocompatibility complex (MHC for short).

Not long before Doug and I got together to brood about the problem, another immunologist, Dick Dutton, had asked me to give a talk to his seminar group about the evolution of the immune system. I chose to attack the question of the evolution of the MHC complex, a question that has exercised the best talents of immunologists for decades. Much is known about MHC, and there is no doubt that our ability to fight off so many diseases has a direct connection with this complicated cluster of genes.

All this work, however, has led to considerable puzzles. In spite of a great deal of effort, it has been very difficult to find a connection between particular MHC alleles and resistance or susceptibility to particular infectious diseases. Typical is a recent observation, made by a group of scientists at Oxford, who found what seems to be a connection between two kinds of MHC genes and resistance to falciparum malaria in a Gambian population. This result was soon cast into doubt when these same scientists looked for this relationship in East Africa and did not find it.

If, as we and many other people suspected, MHC is important in determining resistance and susceptibility to disease, then why isn't the situation more clear-cut? In a straightforward world, possession of one MHC allele would make you resistant to tuberculosis, while possession of another would ensure that you would catch the disease. Obviously, the world of MHC is more devious than that.

In spite of the fact that it has not been possible to find clear connections between MHC alleles and specific infectious diseases, there is no doubt that the MHC genes are very important. One measure of their importance is that many of their polymorphisms are very old. Some of them have been traced back a hundred million years or so. Like jugglers able to keep many balls in the air at once, our ancestors somehow survived during all these millions of generations without losing their MHC alleles. It is rather as if we were to come back to the earth a hundred million years from now, to discover that even after this huge lapse of time our remote descendants are still polymorphic for the A, B and O blood types.

How can we tell that these alleles have had a long history? Alleles of an MHC gene taken from the same human population can differ at a hundred places in their DNA. This could not have happened unless they had pursued separate evolutionary careers for very long periods of time, gradually accumulating a large number of differences. The great age of these alleles tells us that it *must* be important to retain this genetic variation in our population.

Population geneticists have proposed a simple explanation for MHC. Perhaps each allele really does, in spite of all the evidence to the contrary, confer resistance to some specific disease or diseases. If you are homozygous for that allele – that is, if you have inherited one copy of it from your mother and another from your father – then you will of course be resistant to those diseases. But if you are heterozygous, that is, if you have inherited one allele from one parent and another allele from the other, then you might be resistant to twice as many diseases.

But, if this is so, then why after all this time is each of us still limited to a maximum of two alleles at each of these MHC genes? There is no obvious reason why we should be. During our evolution a process called gene duplication has often occurred, in the course of which new copies of many different genes have been inserted elsewhere in our chromosomes. If this has happened for other genes, then why haven't we accumulated extra MHC genes, so that each of us can have the benefit of many different kinds of MHC alleles simultaneously? Then perhaps we could all be protected against every imaginable disease!

A clue to this puzzle, as well as to the even greater puzzle of the lack of obvious connection between MHC and specific diseases, can be found in the workings of the MHC complex itself. These genes do all sorts of things, some of which seem to be in direct conflict with each other. The MHC complex is a set of genes at war with itself.

MHC molecules, in the shape of tiny pairs of jaws, are found on the surface of some of your cells, particularly phagocytes. In the last chapter we met one class of these MHC molecules when we looked at how virus-infected cells can be recognized by your immune system. You will remember that the jaws of these molecules are able to pick up, and present to the outside world, little bits of proteins from invading pathogens that the phagocytes have engulfed in the process of patrolling your body.

These molecular jaws, along with the protein fragments that they have in their firm grip, are then encountered by helper and killer T-cells. When a CD4 receptor on the surface of one of the helper T-cells meets a combination of jaws and protein fragment that it happens to recognize, this sets in train a series of events that causes your immune system to start making the appropriate antibodies. It is this process that can be short-circuited by the AIDS virus, which forces the T-cells to destroy their CD4 receptors.

So far so good. Presumably, the more forms of these genes that you have, the more little bits of foreign protein you are able to present to your immune system. This might explain why there are so many different alleles of these MHC genes – but it still does not explain the nagging question of why you have only a maximum of two alleles at each one. What prevented your ancestors, in the course of evolution, from accumulating large numbers of MHC genes, which could have presented immense numbers of bits of foreign proteins to their immune systems?

The answer to this question, though Doug and I did not realize it at the time, goes to the very heart of our understanding of disease resistance and susceptibility. The MHC complex does not simply present foreign protein fragments to your T-cells. It is also a kind of tutor, and is involved in the very shaping of your immune system. Like any good tutor, it teaches your T-cells the difference between right and wrong.

Right and wrong in this case refers to their ability to distinguish between 'self' and 'non-self'. This tutoring process began when you were very young, still protected by antibodies from your mother but unable to make antibodies of your own. It was during this formative period that another class of genes of the MHC complex played a central role. These genes, too, are very polymorphic. But instead of presenting bits of foreign protein from invading pathogens, they presented to your T-cells little protein fragments that were derived *from your own cells*.

Those molecular tutors were mercilessly strict – if a T-cell incautiously responded to one of those little tidbits of 'self' so temptingly offered, it was promptly destroyed. Unsurprisingly, as the process continued, this strict disciplinarian approach modified your population of T-cells dramatically. The T-cells that were left behind were those that were unable to recognize proteins from your own body. Much of this happened when you were an infant, with the result that now this carefully culled population of T-cells is made up only of those that can recognize the non-self proteins from the various pathogens that now invade your body.

Intriguingly, these two classes of MHC genes work at cross-purposes to each other. If the tutor class molecules presented too many different kinds of 'self' protein to your immune system, they would have killed too many of your T-cells. Then later, when the other class molecules begin to present bits of invading pathogen, they would do so in vain, for you would have no T-cells left that are able to recognize them. This would increase your risk of infection.

On the other hand, if the tutor molecules did a poor job, and presented too few little bits of 'self' protein to your T-cells in the course of their tutoring activities, then they might not have rooted out all the T-cells that recognize some molecules that actually belong to your body. Then, if some MHC molecules were later to accidentally present bits of self protein to this population of insufficiently tutored T-cells, this would trigger the production of antibodies to your own proteins. This kind of mistake is not some theoretical possibility – it often happens. Recognition of self as foreign can literally make you allergic to your own body. It can produce an eventually fatal autoimmune disease such as lupus, or can give rise to the slow and painful degeneration of various forms of arthritis.

This collision of purposes may help to explain why you have relatively

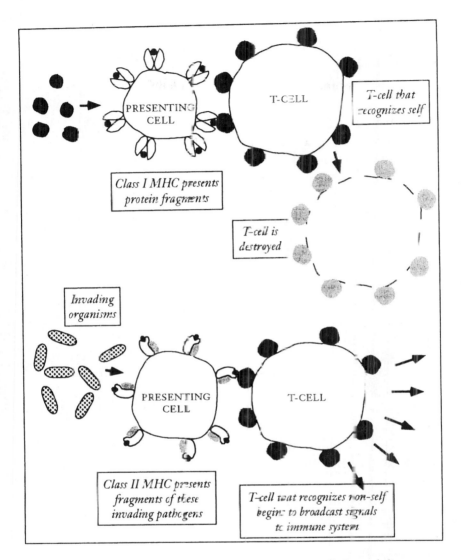

FIG. II-I How the MHC helps the immune system to distinguish between self and non-self molecules

few MHC genes. If you were protected against all possible non-self proteins, you would be unable to recognize self, and your immune system would quickly destroy the tissues of your own body.

Genetic herd-immunity

Doug and I were faced with the task of trying to make sense out of all these observations. To begin with, we knew that plagues are unusual events, even though a disproportionate amount of attention is paid to them. Most of the organisms that live in our guts, our mouths, our lungs and the various interstices of our bodies coexist with us in a cautious *modus vivendi*, almost never becoming too numerous or too widespread. Might the MHC, we wondered, have more to do with controlling this penumbra of potential disease organisms than it does with controlling plagues? It is in building up such complex interactions, after all, that natural selection spends most of its time. This same conclusion was reached independently at the same time by the immunologists Jan Klein and Colm O'hUigin.

And because there are so many such organisms, and their interactions with us are so complicated, it is hardly surprising that it has been frustratingly difficult to find direct connections between particular MHC alleles and the obvious diseases that cause plagues. Protecting us from plagues is not likely to be the primary task of MHC.

Such mild interactions between our genotypes and disease organisms can shape our lives in subtle ways. I recently had experience with such an interaction.

For years I suffered intermittently from a duodenal ulcer, which my doctor attributed to stress, and which he controlled with the standard treatment of histamine receptor blockers that reduced my stomach's production of hydrochloric acid. I was resigned to swallowing these expensive pills indefinitely when, along with millions of others, I read of an amazing discovery made by Barry Marshall, a doctor working in Perth, Australia.

In 1982, Marshall was struck by a remarkable correlation between such ulcers and the presence of a spiral-shaped bacterium, then classified as a *Campylobacter* and since renamed *Helicobacter pylori*. Could the bacterium, rather than stress, be the cause of these ulcers? His co-workers patiently pointed out to him that correlation does not imply causation, and that there was no reason to abandon the received wisdom that ulcers are caused by stress. But Marshall was convinced that this correlation was not coincidental. Finally, in July 1984 and in the grand tradition of the great microbe hunters of the past, he performed a dramatic experiment. He swallowed a culture of *Helicobacter* and waited to see what would happen.

Within days, he was in pain. A probe with an endoscope showed that the lining of his stomach and upper duodenum had broken out into dozens of crater-shaped sores. His body soon fought off the infection, but he had proved his point. Our stomachs, long thought to be completely sterile, can readily harbour these bacteria, which can survive the acid conditions by hiding under a layer of mucus secreted by the stomach cells.

Marshall, who has now moved to the University of Virginia, currently spearheads an intensive research programme investigating the bacillus. It took many more years of studies and tests before the majority of the medical establishment began to be convinced. The drug companies that market the receptor blockers, of course, were less than pleased, but they made no overt attempt to suppress Marshall's results. On the other hand, they have made no particular effort to publicize them.

Were these unwelcome creatures accompanying me on my voyage through life? The answer was yes. The presence of *Helicobacter* was soon betrayed by a simple metabolic test (though I had some trouble persuading my doctor to do it), and the bacteria have now been banished from my stomach by a short treatment with a combination of antibiotics.

But what did my genes have to do with it? Quite a lot, it appears. You will recall that I am blood type O. *Helicobacter*, it turns out, binds most strongly to the distinctive carbohydrate chains on the surface of the cells of people who have this blood type, and less strongly to those of people with types A and B. This explains a puzzling observation that has been found repeatedly in many (though not all) studies of the association of blood types with ulcers during the last half-century. People with blood type O are more likely to get stomach ulcers than those with other blood types.

My wife and daughter, protected from a similar infection by their B alleles, can munch their way through an assortment of highly spiced ethnic foods with an insouciance that, until recently, I could only envy. I have a lot of catching up to do.

None the less, there are many people with blood types other than O who suffer from *Helicobacter* infections. The protection conferred by these alleles is not perfect. And *Helicobacter* is now being implicated in a great variety of other diseases, including stomach cancer and most recently heart disease. It is also now emerging that this bacterium, like so many other infectious agents, is much more harmful in the Third World. Children in West Africa who are not breast-fed, and who consequently do not receive the protective benefits of their mothers' antibodies, show a much higher level of *Helicobacter* infection than those who do get antibody-laden milk from their mothers. And here is where the game turns deadly. For *Helicobacter* can gnaw holes in the protective lining of these children's stomachs, so that other even more harmful enteric bacteria that are everywhere in their environment can invade. These invasions can trigger further damage and even septicaemia. *Helicobacter* does not just cause mild health problems in nervous professors and highly-strung corporate executives. It is an important part of the penumbra of diseases that surround us, and like so many of the members of that penumbra it does its greatest damage among the poor and in the tropics.

How do we protect ourselves against this ceaseless drumbeat of infection? If it is our genetic diversity that does it, then is MHC the main source of that diversity?

Doug and I needed one final piece of the puzzle before the story began to fall into place. It has been known for decades that the MHC type of a donated organ is very important in determining the speed at which that organ will be rejected. If the MHC alleles of the person who donates the organ are very similar to those carried by the recipient, rejection will be, although not averted, at least delayed.

The hope of understanding MHC's role in tissue rejection has driven much of the research into this gene complex. Surely, if MHC could be understood, then the process of tissue rejection could be controlled. Yet, as MHC researcher Walter Bodmer remarked to me some years ago, there is a 'dirty little secret' about MHC that tends to be downplayed by scientists working in the area. No matter how carefully donor and recipient are matched for MHC, the donor tissue is still eventually rejected unless immunosuppressive drugs are used at the same time. Completely successful transplants can be made between identical twins, but of course twins are matched, not only for their MHC types, but for all the other genes in their genomes.

The situation is actually worse than I have painted it. While there is some correlation between the extent of a match and the speed of rejection, it is not a particularly close one unless donor and recipient happen to be related, preferably if they are siblings. No matter how carefully the match is made between the MHC types of two unrelated people, the tissue is rejected rapidly. If an equally careful match is made between two related people, rejection is less rapid. Bodmer and other scientists assumed that this was because there were unknown genes in the MHC complex that could not be matched but which helped to control the rejection process. Such genes would tend to be different in unrelated people and to be the same in related people.

But there was evidence from mouse experiments that genes elsewhere on the chromosomes play a role in tissue rejection, so these genes do not necessarily have to be in the MHC complex itself. It appears that it is the interaction between MHC and all these other genes that results, not only in the unique shape of our immune systems, but the unique shape of the genetic 'fingerprint' that we all carry on the surfaces of our cells. Unless we happen to have an identical twin, this fingerprint is unique to each of us.

The final shape of our immune system is the result of tutelage by MHC molecules, but these molecules must of course work with what they are given – in this case the fragments of 'self' protein that have been derived from the other proteins of our bodies. And, because we are so genetically diverse, that set of protein fragments is unique to each of us. Suppose it were feasible to transplant your entire MHC complex into the sixth chromosome of an unrelated developing human foetus, replacing the foetus's original complex. It would be found that after the transplant, and after the foetus had grown to maturity, the resulting adult would still be able to reject your tissues even though he or she carried your MHC complex. Experiments with mice, though not quite as cleanly designed, have given just this result. The way an individual's immune system is shaped, and the suite of proteins that it recognizes as non-self, is the result of an interaction between MHC and the rest of the individual's genes. As a result, the immune system of each of us has a unique set of capabilities.

Subsequent modification of the immune system by the activities of the MHC molecules which present bits of pathogen seems to have little or no effect on tissue acceptance or rejection. Each of a pair of identical twins is inevitably exposed to a somewhat different mix of pathogens during the time that they are growing up, and yet each can still readily accept tissue from the other twin. So our specific set of defences is created early, during the time that our immune system is maturing and the tutor MHC molecules are shaping it.

Of course, this immensely complicated system is not perfect, and there are many compromises built into it. It is easy for tutor MHC molecules to teach your immune system that a protein in your body is 'self' if that protein happens to be present in large amounts, but much harder if it is made in small quantities. This can be demonstrated experimentally. Rare 'self' proteins can be isolated from mice, manufactured in the laboratory in large amounts, and then injected in far greater than normal quantities into the same strain of mice from which they were originally derived. It is often found that these self proteins can produce an antigenic reaction in the mice, just as if they were non-self. It seems that our immune systems can still function well, and effectively ward off many different diseases, even though they cannot sort out more than a minority of our tens of thousands of different proteins into self and non-self.

In spite of these compromises, each of us presents a unique problem to the pathogens that invade us. The immune system of each of us is different from those of our neighbours, not only because we differ in the genes of the MHC complex, but because we differ in many other genes as well. This is unavoidable, and is a consequence of the very genetic diversity that helps to make us such an interesting species. But what, Doug and I then asked, are the consequences to the pathogens themselves, when they are faced with this bewildering variety of defences?

There are several routes open to the pathogens. First, they might find a way to circumvent their host's immune system, as the syphilis spirochete seems to have done. Some of them, like the spirochete, coat themselves with host proteins as a disguise. Others, like the malaria parasites, hide themselves in the host cells where the immune system cannot reach them. The AIDS virus is the only pathogen we know of that can destroy its host's immune system permanently – but many others, like the bacteria that cause *verruga peruiana*, can depress their hosts' immune systems temporarily to dangerously low levels. Yet for many of these pathogens, as we have seen, there is a trade-off – in the process of disguising themselves, they may lose their ability to live outside their hosts, or they may become so fragile that they cannot easily be transmitted from host to host.

A second possibility is that the pathogen can break free of its normal route of transmission and take an entirely different route. This is what the plague bacillus has done. Here too there can be a trade-off – the plague bacillus puts its host in harm's way from the result of its uncontrolled multiplication.

A third possibility, and the most interesting one from the standpoint of genetics, is that the pathogens can specialize. Faced with such a bristling variety of host defences, pathogens might concentrate on the members of the host population who happen to be most susceptible to them.

This last group of pathogens, we thought, should have interesting properties. Because of their specialization they will only be able to affect a minority of the host population. If that minority happens to be dispersed throughout a host population that is generally resistant, the pathogens will have great trouble spreading from one susceptible host to another. Perhaps they will be forced to hide in some out-of-the-way cranny in the body, like the typhoid bacillus. Or perhaps they will spend most of their time infecting an alternative host, like the virus that causes influenza and that does most of its evolving in the pigs of southern China.

Of course, if the minority of susceptible hosts were to increase in numbers for some reason, then the pathogens that specialize on them could also begin to spread. Once this happens, it would not be long before these susceptible hosts would begin to fall victim to the pathogens. The hosts would once more become rare, and this in turn would force the pathogens to beat a retreat.

Suppose, we thought, there were many such specialized pathogens affecting our species. Then the frequencies of the diverse genotypes in our population, and the numbers of all these pathogens that affect us, would be in a continual, dynamic flux. Host genotypes susceptible to a particular specialist pathogen would be kept in check by that pathogen, and the pathogen would in turn be kept in check by the relative rarity of the hosts that it could easily attack. This in its essence is genetic herd-immunity.

In the genetic herd-immunity model, there is safety in diversity. Suppose that you are susceptible to disease A but resistant to disease

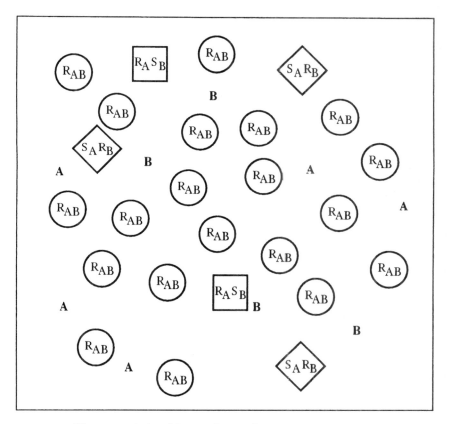

FIG. 11-2 How genetic herd-immunity works

B, while I am susceptible to disease B and resistant to disease A. Suppose further that I am part of the majority of the population that can resist the onslaughts of disease A. My presence, and the presence of other members of this majority all around you, ensures that the pathogens causing disease A cannot multiply to high numbers, and this of course helps to protect you. You, on the other hand, are part of the majority of the population that is resistant to disease B, and this majority to which you belong protects me by not permitting the pathogens causing disease B to multiply unchecked (Figure 11-2).

In the figure, the majority of the hosts, represented by circles, are resistant to both diseases A and B. But there is a minority, represented by squares, who are resistant to A and susceptible to B, and another minority, represented by diamonds, who are susceptible to A and resistant to B. (Some unlucky hosts, of course, are susceptible to both diseases, but they are very rare.) A few pathogenic organisms of both types, represented by the boldface letters A and B, are scattered thinly through the population, but because each has so few susceptible hosts to prey on neither pathogen can become common. In the lower part of the figure a pathogen of type B has come in contact with a host susceptible to disease B, but you can see that such an event is rather unlikely. However, if hosts susceptible to disease B should become more common, the pathogens of type B will soon multiply as well, and the chance of such an interaction will increase. Such hosts will quickly be reduced in number and the pathogens will soon follow suit. The same pattern will occur with pathogens of disease A. Neither host genetic variant will be lost from the population.

Now imagine that diseases C, D and so on are also loose in the population, and that a minority of the hosts resistant to A and B are susceptible to these other diseases. These diseases, as well will be kept at low levels in such a polymorphic host population. So, perhaps, will many others.

Our model, we soon realized, makes some interesting predictions. One is that there should be selection for host diversity. There certainly is lots of diversity in the human population – you will remember all those MHC alleles that are so different from each other and that have managed to survive in our ancestors for tens of millions of years. Perhaps genetic herd-immunity is one reason why we are so diverse?

The second prediction is that there should be selection, over long spans of time, for a high degree of pathogen specialization. In the normal course of events, if the hosts are highly diverse, then no one pathogen can be versatile enough to attack all the members of the host population. The only recourse for the pathogen is to do a really excellent job of attacking the minority that happens to be most susceptible to it.

The third prediction follows from all the possible interactions between our immune systems, our MHC alleles, and all the rest of our genes. Because of all this complexity, we can hardly expect a clear-cut relationship between our possession of a particular MHC allele and our resistance or susceptibility to a particular disease. It is exactly the lack of such a relationship that has confused and frustrated generations of immunologists who have tried to make such connections and almost always failed. If the model that Doug and I constructed is right, while there is indeed a connection between genetic variation and diseases, the connection is so complex that it will take generations of scientists to unravel it.

The fourth prediction is that this herd-immunity balance should be an exceedingly delicate one. It can easily be upset. A host population in a region free of a particular disease might by chance lose some critical genes and become universally susceptible to that disease, so that when it is introduced most of the hosts succumb. *Helicobacter* infections are very common in Peru, where they are spread through the water supply. Virtually all native American Indians are of blood type O, and most Peruvians are of Indian ancestry. We do not know of course at what point *Helicobacter* was introduced into the New World – it might have come over with the early migrations across the Bering land bridge from Asia. But if it was introduced later by the European explorers, perhaps this explains why it is now so widespread among native Americans, since they have lost the protective A and B alleles. And perhaps, although Yale's Francis Black disagrees with me on this point, other genes for susceptibility, which have accidentally spread through the native American population, may help to explain their extreme sensitivity to introduced diseases – remember the violent course that vivax malaria took among Peruvian highlanders when they descended into the malarious rainforest.

The herd-immunity balance can also be upset by the appearance of a plague. With plagues, all bets are off. Most of us are susceptible to plague organisms, which by sheer numbers or sheer virulence can overwhelm our defences. (Though not all of us – remember the few who did not succumb to the Plague of Justinian, and the minority of the Nairobi prostitutes who have remained free of the AIDS virus.) Later, when the unusual ecological conditions that sparked the plague disappear, along with the outbreak of plague pathogens, then the survivors and their regular accompanying crowd of specialist pathogens will soon go back to their uneasy and complex *modus vivendi*.

All this is very well, as far as it goes. Building models is lots of fun. The herd-immunity model can be simulated on a computer, and it behaves as expected. It easily permits a genetically diverse host population to keep a great multiplicity of diseases at bay simultaneously, and at the same time selects for even more diversity among hosts and a high degree of specialization among their various parasites. But the scientific literature is filled with lovely models that have turned out to bear no relation to reality. How could Doug and I prove or disprove the genetic herd-immunity model?

The more we thought about it, the more difficult the problem became. Statistical tests alone would be unlikely to give unambiguous results. Perhaps mouse colonies could be infected with assorted diseases, and the genetic consequences measured. But such tests would be expensive, logistically complicated, and unlikely to produce a clear answer. This is because the diseases most likely to contribute to genetic herd-immunity are also, frustratingly, the ones least likely to cause detectable damage among their hosts. Diseases that are virulent enough to have noticeable effects probably operate outside this complex genetic balancing act. And, of course, the custodians of mouse colonies are understandably less than enthusiastic about researchers who suddenly turn up and let diseased mice loose among them.

There is also one obvious and very large missing link in our chain of evidence. Examples of pathogens that are known to be specialized to specific subsets of a host population are few and far between. Yet the model requires that these pathogens be numerous. How could we look for them? Perhaps we could examine the different mixes of potentially pathogenic bacteria in the guts of different people, but the results would be more likely to be affected by what our subjects had eaten for dinner than anything else. We were frustrated at every turn.

Then I happened to take a trip to the rainforest.

12

Why so many diseases?

... [C]an it be seriously suggested that a rather uniform area of Amazonian rain forest provides, in 3.5
hectares of land, anything like 179 separate ecological
niches for trees? ... Pest pressure is the inevitable,
ubiquitous factor in evolution which makes for an
apparently pointless multiplicity of species in all areas
in which it has time to operate.

J. B. Gillett, 'Pest pressure, an underestimated
factor in evolution', 1962.

Why is our species afflicted by so many diseases? And why do these
diseases swarm with such diversity in the tropics and not in the temperate zones? There are trivial answers to these questions, of course. The
tropics are hot and humid, which allows disease organisms to multiply
swiftly, so that they can easily contaminate food and water supplies.
Because of the plentiful moisture everywhere, the fragile parasites are
less likely to die from desiccation when they are exposed outside the
body of their host. Further, they can be passed easily from one host to
another on films of water or sweat. Seasons are not pronounced, so the
parasites are not killed off by cold weather. And because of the sheer
abundance of life there are far more possible alternative vectors for
disease, ranging from mosquitoes and biting flies through ticks to fresh-
water snails. As a consequence, diseases in the tropics do not need to
be as sophisticated as those in temperate zones.

Because it is not as demanding to be a tropical disease, the tropics
provide plentiful opportunities for disease organisms that would have
little chance in the temperate zones. We have already seen many

instances of this. For example, falciparum malaria, with a less complex life cycle than vivax or malariae malaria, does very poorly away from the tropics and the subtropics.

But there are deeper answers as well. To really understand the differences between the temperate zones and the tropics, we must understand tropical diversity as a whole, and examine the forces that have driven its evolution. Remarkably, diseases themselves are likely to have played a large role in the generation of the immense variety of plant and animal species that populate the tropics, and this growing variety of species has in turn driven the evolution of the diversity of diseases that prey on them. Diseases form an important part of the story of life on our planet – and, frightful though many of them are, the world would be a far less rich place without them.

All this was vividly brought home to me in the Amazonian rainforest. The forest itself seems like an unlikely place to learn something about the diversity of human diseases, and the connection between the two may not be obvious at first. But I beg your patience. An understanding of the forces that have shaped the evolution of the rainforest will, by the end of this chapter, suggest how genetic herd-immunity might operate to help produce great genetic diversity in our own species. The intellectual journey that follows is, with all its detours, well worth making.

The sound of many trees falling

Manù National Park, half the size of Switzerland, encompasses the entire drainage basin of the Manù River in southern Peru. The river gathers water from the rain-soaked cloud forests of the eastern slopes of the Andes and eventually channels it into the Amazon. The numerous tributaries of the Manù, fed by torrential rains, tumble briskly down the Andean foothills, then slow as the terrain flattens, and finally begin to wind in leisurely fashion through the lowland rainforests.

If you could speed up time and watch the Manù River and its tributaries over a span of thousands of years, you would see them writhe lazily like snakes. The outer banks of the sinuous bends in the river are eroded away during the pell-mell rush of the rainy season, to the extent of several metres a year, and the soil is washed clean and deposited as great sandbanks on the inner banks. Sometimes this erosion punches through the neck of an entire sinuous curve and the river, suddenly straightened, rushes past, leaving behind an oxbow lake that is isolated and becalmed in the forest. It is not long before the forest reclaims the lake, as exuberant tree growth fills it in from the margins. This new patch of forest continues to change, slowly accumulating a great diversity of trees until, over a span of hundreds and perhaps thousands of years, it becomes indistinguishable from the mature rainforest that surrounds it.

As they wander widely over the landscape, the rivers too are continually renewing it. On the broadening sandbanks, brief growths of shrubby *Tesseria* and dense stands of *caña brava* are soon replaced by fast-growing *Cecropia* – trees that race towards the light, grow in segments like bamboo, and increase their diameters by an astonishing 20 centimetres a year. These in turn are replaced in a decade or two by many other species of tree, specialized to grow to different heights and to utilize different amounts of the sunlight that filters down to the forest floor. Gradually the diversity of the forest increases, until in a square a hundred metres each side there may be 300 or more species of tree – compared with at most twenty in a similar area of forest in the planet's temperate zones.

It may take hundreds or even thousands of years to reach this point, and during all this time the numbers of mammals, insects, small epiphytic and parasitic plants, and micro-organisms that come to inhabit this maturing region of the forest are increasing as well. In Manù this process has been taken to its ultimate, for it is now one of the most diverse terrestrial ecosystems on the planet.

Even this highly diverse rainforest is continually changing, as trees occasionally fall to the ground, leaving holes in the canopy that are frantically exploited by desperate saplings striving for the sun. During the time that I was in Manù there was a mild windstorm that lasted an hour or so, with a scattering of rain. The gusts of wind could charitably be called no more than brisk. None the less, the effects were astonishing. We saw two trees fall on the edge of the forest during the course of the storm. Many other trees fell in the forest's depths, although – and this would have pleased Bishop Berkeley – their effects were mostly invisible to the casual visitor. Later, as we ventured along the trails after

FIG. 12-1 The Manù rainforest

FIG. 12-2 Treefall after a windstorm in Manù

the storm, we found that dozens of trees had crashed down and blocked them at various points.

Such freshly-fallen trees always provide a marvellous opportunity for the naturalist, bringing samples of the teeming life of the canopy down to ground level. We spent a good deal of time examining this wealth of life. I noticed that many of the broken trunks of the trees that we clambered through and around had extensive signs of rot, though others seemed healthy enough. The crowns of the rotted trees, in particular, were loaded with thousands of epiphytes, the weight of which had probably helped to bring them down.

Since so many trees fell during this little storm, one might expect the forest floor to be littered with the remains of older trees that had fallen victim to earlier storms. In the temperate rainforests of the Pacific Northwest there is just such a tangle, like a giant version of the children's game pick-up-sticks. Yet in Manù there are very few logs and fallen trees, and those few have already begun to rot, quickly sprouting an astonishing variety of shelf fungi. With the help of termites and swarming beetles, these and other fungi quickly break down the wood of even the largest tree. The simpler nutrients that are produced by these forces of decay are rapidly taken up by the living forest's web of shallow roots.

Sometimes these forces of destruction do not even wait for the tree to fall. At one point we found a thirty-foot tree-trunk jutting up in the forest, with all its branches gone, and covered with the dark bulges of termite nests and a maze of mud-roofed little runways. The termites were so active that a mix of their droppings and small wood fragments was visibly sifting down, forming a large pile of sawdust at the trunk's base. The tree was literally being eaten away before our eyes.

There is no depth of soil in the rainforest, since the nutrients are so quickly recycled, which is one reason that trees tend to fall so readily. But the primary reason for this rapid turnover is that the exuberant life of the rainforest is counterbalanced almost exactly by an equally exuberant death. And it is the death as much as the life that determines the rainforest's diversity.

A diversity of trees, a diversity of diseases

A walk through the mature rainforest of Manú is a breathtaking experience. Every tree seems to be different, and is encrusted with a different mix of lichens, mosses and algal mats that leap out in complex detail when observed through a hand lens. Virtually all the trees are hosts to hundreds of different species of bromeliad, orchid and even cactus, growing in profusion high up in the canopy. These epiphytes are hosts in their turn to a great variety of smaller plants and insects. In addition, many trees have evolved elaborate means of protection. Some have intimate associations with ants that provide protection in return for food and shelter. Others are protected by stingless bees that build nests in hollows of their trunks – stingless they may be, but they can bite fiercely. This protection, however, tends to be only temporary. While young *Cecropia* trees are protected by ants that live in the bamboo-like hollows of their trunks, as the trees mature the ants disappear, leaving them prey to a variety of diseases and predators.

Sometimes there are so many complex interactions between the trees and the creatures associated with them that it is impossible to count them. The strangler fig is a good example. Young strangler figs start life as apparently innocent epiphytes, growing in a bit of soil that happens to have accumulated in a crevice high in a tree. They send down delicate aerial roots to the floor of the forest far below, roots that grow thicker with time and begin to twine around the tree that supports them. Eventually the host tree, unable to grow further, dies and rots away in the remorseless embrace of the fig. The parasite now becomes a massive tree in its own right, towering triumphantly, taller than its victim, its trunk made up of a knotted maze of these fused aerial roots.

All during this time the fig itself plays host to a growing range of other animals and plants that come to live within the interstices of the roots as they thicken. A naturalist could spend a lifetime examining a mature strangler fig and still not find all the creatures that have come to live in the innumerable crevices of its trunk.

Tropical ecologists have observed all this diversity with astonishment for nearly two centuries, ever since Alexander von Humboldt became the first scientist to penetrate into the depths of the Orinocan rainforest of the north-east Amazon basin. The more closely the rainforest is examined, the more variety is revealed. And the rainforest has changed our whole concept of the diversity of life on the planet.

Up until recently, cautious zoologists and botanists had simply extrapolated from the million or so known and described species of animals and plants, in order to make estimates of the probable total number of species on the earth. It was thought that there might be three million or so. But recently, thorough censuses have been carried out on a few samples of the epiphyte populations of the rainforest canopy. The results have caused entomologist Terry Erwin of the Smithsonian's National Museum of Natural History to raise the estimate of the number of species on our planet from three million to a dizzying ten to thirty million, the majority of which live in the world's rainforests. Why this insanely exuberant variety? Why so many species, when a temperate forest has far fewer?

The question, complex as it seems, really boils down to the problem of what maintains the diversity of the trees themselves. Alfred Russel Wallace, who along with Charles Darwin suggested the idea of natural selection, also observed something remarkable about the distribution of these trees. In *Tropical Nature* (1878) he wrote:

> If the traveler [to the rainforest] notices any particular species and wishes to find more like it, he may often turn his eyes in vain in every direction. Trees of varied forms, dimensions and colours are all around him, but he rarely sees any of them repeated ... He may at length, perhaps, meet with a second specimen a half a mile off, or he may fail altogether, till on another occasion he stumbles on one by accident.

In a mature rainforest, as Wallace observed, the diversity of tree species is so great that two trees of the same species seem almost never to be found growing next to each other. This great variety of tree species can in turn support a great variety of other animals and plants.

As long as the trees themselves are diverse enough, they provide a kind of motive force for further evolution. The bewildering variety of animals and plants that we see in the rainforest will inevitably evolve and be maintained if the trees on which they depend are very different from each other. The many different species of trees form a structural underpinning that supports the whole complex ecosystem.

Once this is lost the whole magnificent edifice topples. A tropical

forest ecosystem is unexpectedly, even frighteningly, fragile. To study
the impact of deforestation, Thomas Lovejoy, an ecologist at the Smithsonian Institution, took advantage of a Brazilian law designed to help
protect the rainforest. The law requires that fifty per cent of a forested
area scheduled for clearing should remain intact. Lovejoy and his colleagues persuaded farmers who were clearing new land in a region just
north of the Amazonian city of Manaus to leave untouched forested
patches of different sizes, ranging from a hundredth of a square kilometre to two square kilometres. They found that while the smallest
patches showed the greatest disturbance, even the largest patches lost
many species. An 'edge effect' was particularly striking – when previously intact forest was exposed to the drying effects of the wind,
many trees died and many species of animal, particularly large ones,
disappeared.

But in an intact forest each species of tree – and indeed each individual
tree – provides its own set of ecological riches for the other creatures
that come to inhabit it. Thus, as the diversity of tree species increases
in the course of evolution, the ecological opportunities for the many
other animals and plants that live in association with them are enormously multiplied. All this is plain enough, and it is easy to see how a
rainforest tree like a strangler fig provides such a great variety of ecological niches for all the creatures that come to inhabit it. But what
kind of ecological niches do the trees themselves fit into?

Scientists have been fighting over the definition of an ecological niche
ever since the pioneering British ecologist Charles Elton proposed the
term in the 1930s. Elton supposed them to be what their name sounds
like, little spaces of unusual environment in which a particular species
can always outcompete other species, even similar ones. In his view of
the world each species occupies its own niche. A species is rather like
the statue of a saint that occupies each of the niches in the buttresses
of the nave of the cathedral at Chartres, each one excluding by its
presence in that niche all the other saints in the pantheon.

The great puzzle is that in a tropical rainforest there seem to be too
many kinds of tree for the few ecological niches that are available to
them. It is not easy to see how these hundreds of species of tree in
the rainforest can somehow all be occupying distinctly different little
niches.

Certainly there are a variety of obvious niches in the rainforest. In the humid tropics all the trees are competing for the intense sunlight that burns down from the equatorial sky. So effective is this competition that very little sunlight manages to penetrate to the forest floor. But if a tree falls and opens up a hole in the dense canopy, sunlight can flood in. The saplings that can most swiftly exploit this opportunity are the ones that can, at least in the short term, utilize this sudden burst of light.

Ecologists have agonized over what is likely to happen next. Perhaps certain species should preferentially grow in the gaps, gradually spreading through the forest as more gaps appear here and there. If this is indeed what happens, and these trees live for a long time, then this would lead over time to an ever-shrinking diversity, as the trees that are most successful at exploiting gaps edge out the less successful ones. Indeed, mathematical studies have shown that if many different species are competing for a single resource, the most successful one should inevitably replace the others.

Alternatively, the diversity of light intensities in a gap may provide a great diversity of niches, so that many different species of saplings will start to grow, each specialized to take advantage of different amounts of light. The number of species of tree that fill the gaps might actually be greater than the number in the surrounding forest.

If this is so, then these gaps might be the secret to preserving rainforest diversity. The great diversity of trees growing in the gaps would continually generate little islands that contain a larger variety of species than is found in similar-sized areas of the forest as a whole. Over time, as the gaps in the canopy close again, these islands of high diversity would gradually dilute out into the lesser but still considerable diversity of the mature rainforest.

On the face of it, this idea does not seem to be very likely. Small variations in the amount of sunlight are surely not important enough to maintain, much less to generate through natural selection, all the different species of tree in the forest, with their myriad growth habits, shape, biochemistry, reproduction, and so on. The fundamental problem is that physical conditions really do not vary a great deal in the rainforest, even in the gaps.

None the less, the possibility has to be tested. To do it adequately, one must learn an incredible amount about the dynamics of gaps in the rainforest, and this in turn requires a really good data base. Stephen

Hubbell, an ecologist at Princeton University, has spent fifteen years building up exactly such a data base. With the aid of a large group of enthusiastic amateur and professional ecologists, he has examined a fifty-hectare plot of mature high-canopy rainforest on Barro Colorado Island, Panama, in astonishing detail.

In the course of their work they had to contend with the forest's innumerable denizens, including venomous stinging ants like the fierce *Odontomachus* and the glossy, three-centimetre-long *Paraponera*. The latter has the endearing habit of dropping down into the space between an ecologist's neck and shirt, and its sting can cause agonizing pain for several days. They also had to do their best to avoid touching the temptingly beautiful long silky hairs of megalopygid caterpillars, the delicate poisonous tips of which can break off and work their way into the skin, causing disabling pain for weeks. And they worried a great deal about venomous snakes like the fer-de-lance and the bushmaster, which luckily are relatively rare in the mature rainforest.

Braving these and other discomforts, this army of workers managed to identify almost a quarter of a million trees of more than three hundred species in this small area, and to plot their positions carefully on a computerized map. This huge body of data allows changes in the forest to be examined over time in great detail.

Using this map, Hubbell and his co-workers were able to follow, over a period of eight years, how the gaps that appeared as trees fell were filled in. Their study, just completed, has been a surprise to many ecologists. It turns out that there is just as much diversity, no more and no less, among the trees that colonize the gaps as there is in the rest of the rainforest. Whatever forces are maintaining rainforest diversity, then, do not seem to be directly connected to some unusual property of the gaps. If there really are more ecological niches in the rainforest than are immediately apparent, then we must look elsewhere for them.

Do rubber trees cast a shadow?

The rubber tree, *Hevea brasiliensis*, native to the Amazon, is one of thousands of species of plant that make a white syrupy sap called latex. It happens to make far more of this sap than most of its relatives, so that the latex bleeds out in copious quantities when the tree's bark is cut. There are many *Hevea* trees in the lowland rainforests of the central Amazon, but like most other trees of the mature rainforest they are almost never found in groups. Instead they are scattered thinly through the forest.

Before the Europeans arrived, the inhabitants of the region used to make little waterproof bags from the dried latex. This unusually elastic material, brought back by early explorers, became a curiosity in eighteenth-century Europe. But it remained an amusing novelty with no practical uses, for it tended to melt whenever the temperature rose. It was not until 1839 that latex finally became a commercially viable product. In that year, after much experimentation, the American inventor Charles Goodyear mixed latex with sulphur and heated the mixture, turning it into a hard resilient black substance that would not turn into a gooey mess whenever the weather got warm. His discovery founded the modern rubber industry.

During the nineteenth century the harvest of latex from these scattered *Hevea* trees of the Amazon brought a ghastly kind of prosperity to the area – a prosperity that was founded on the brutal enslavement of the Indians. Then, in 1876, a Brazilian official was bribed to allow 70,000 seeds to be exported to England, where they were planted in Kew Gardens. Some of the 2800 survivors were sent to Singapore and Ceylon, where they were soon spread by colonial planters to the rest of southern Asia and to western Africa. It was not long before they were thriving in serried ranks in great plantations. (As we have seen, the seeds of brutality and worker exploitation were unfortunately planted along with them.)

The rest is history. In 1910 the trees from the Far East began to produce substantial amounts of easily harvested latex. Within a few years the rubber-based economy of the Amazon collapsed, turning major towns along the river like Manaus and Iquitos into decaying shells of their former selves.

Attempts were soon made to start huge plantations of rubber trees in the Amazon, to try to mimic the success of the plantations of Southeast Asia. The largest was that of Henry Ford, who was busy putting America on wheels and whose demand for rubber was rapidly becoming insatiable. The Brazilian government gave him a thousand square kilometres of land a short distance south of the city of Santarém on the Amazon, which he promptly named Fordlandia. In 1927 his managers began the project, clearing the forest and planting hundreds of thousands of seedlings.

While his managers built houses suitable for Detroit winters in the great openings that were hacked out of the jungle, Ford himself tried to impose his Calvinist ethos on the local population. All this might not have mattered if the trees had thrived, but they did not. When the land was cleared the topsoil quickly baked into hardpan and eroded in the torrential rains. The trees shrivelled in the direct sun. And, most importantly, the young trees succumbed to caterpillars, to a disease caused by the fungus *Microcyclus ulei* that attacked their leaves, and to other fungi that attacked their roots.

Ford's workers tried shading the trees, and even moved the entire plantation to a new area. They also carried out massive grafting of Far Eastern rubber tree saplings, which they thought might be disease-resistant, on to the Brazilian rootstock. At one point, three-layer grafts were tried, with supposedly fungus-resistant roots grafted to supposedly high-yielding stems and supposedly *Microcyclus*-resistant leaves. All their attempts failed. Over three million trees were eventually planted, and yet the yield of latex in the best year was a piffling 700 metric tonnes. Ford finally sold all his land back to the Brazilian government in 1945.

The rubber plantations of the Far East and Africa are not completely free of diseases, because local pathogens have gradually begun to attack them. But they have remained highly productive down to the present time. In 1993 rubber production in South Asia was almost 500,000 metric tonnes, in Southeast Asia 4.5 million metric tonnes, while Brazil only managed to produce 26,000.*

When they are scattered thinly through the Amazonian rainforest, rubber trees thrive, but when they are brought together in groups they succumb to diseases. Is there some clue to rainforest diversity in this story? Might diseases and perhaps larger predators, if they were numerous enough and specific enough, produce the diversity of niches that is required to explain the profusion of rainforest trees?

---

* Very recently, successful and highly productive rubber plantations have been established on a cool dry plateau inland from São Paulo, safely away from the trees' natural predators and diseases.

The ecological chequerboard

In a little-noticed paper published in 1962, Jan B. Gillett, an ecologist working in Kenya, pointed out that predation can actually be a powerful motive force in evolution. This, he suggested, is because it can make the environment more diverse for the species being preyed upon and thus produce an even greater variety of ecological niches.* A few years later, and independently of each other, a pair of tropical ecologists, Joseph Connell of the University of California at Santa Barbara and Daniel Janzen of the University of Michigan, both elaborated on Gillett's suggestion that numerous ecological niches could be generated by disease and predation. They envisaged the following simple scenario.

In a mature tropical rainforest, there are far more diseases of trees – because there are far more species of tree – than the few that attacked Henry Ford's monoculture of *Hevea* trees in his plantations. Many of these must have evolved a high degree of preference for particular species of tree. All these creatures – bark and wood beetles, bacteria, fungi, even viruses – will thrive on the particular kind of tree that they are specialized to attack, and in the soil around it. They will tend to cluster around the tree. Now, suppose further that these parasites tend to be particularly numerous near adult trees (which have managed to grow to be big and strong enough to withstand their depredations). Then any seeds of these trees that fall nearby, or seedlings that begin to germinate from these seeds, will very likely be preyed upon by all these locally numerous disease organisms. They will also be eaten by a variety of seed-eating predators that are attracted by their abundance. These seeds and seedlings will of course be far more vulnerable than the well-established adults, so that their survival rate will be low.

Seeds that happen to fall outside this exclusionary region, however,

---

* Gillett went a step further, and anticipated some of the arguments I made in the previous chapter. In his remarkable paper he pointed out:

'The answer to a problem which has been discussed by [J. B. S.] Haldane, why each species does not come to consist of a single genetic type, that most suited to the environment, is also plain: the effect of pest pressure in producing genetic diversity applies at the intraspecific as well as the specific, genetic and family level. That strain within a species which is commonest is the one to which the pests of that species will tend to become most adapted; therefore the viability of this strain will be reduced and other forms will be favored.'

beyond the reach of the parasites, will be more likely to survive. Eventually, of course, the progeny of the specialist parasites, spreading through the forest, will be sure to discover them. But by the time this happens the seedlings will have become robust enough to be resistant to their more severe effects. The result will be that the parasites can multiply on or around the new tree, but they will not kill it – at least not immediately. Soon a new cluster of parasites will gather in large numbers around the new tree, forming their own *cordon sanitaire*, preventing seeds of the same species from germinating nearby.

If this scenario is correct, then these parasites, specialized to particular tree species, will have the effect of chopping the rainforest up into little islands. They will provide a huge variety of ecological niches that will be quite invisible to the casual observer. This elegant explanation, Janzen and Connell thought, might be the answer to why there are so many tree species. Most of the other tree species could germinate freely within the cluster of parasites, but each would soon be surrounded by its own cluster of specialized predators and parasites. Consequently, trees would be spread out more evenly than you would expect by chance. There would be plenty of room between them for trees of other species, themselves equally spread out through the forest because of the effects of their own accompanying pathogens.

If you could see all the different exclusionary regions of the pathogens involved, they would look rather like a complicated set of overlapping chequerboards of different colours. An elegant picture indeed, but is it really true? Can the Janzen-Connell hypothesis really provide enough niches that are sufficiently different from each other to maintain the diversity of the rainforest and – more importantly – drive the actual evolution of that diversity?

Encouragingly, exactly the kind of spacing between species that the model of Janzen and Connell predicts was discovered by the ecologists Randall Ryti and Ted Case, from my own University of California campus. They discovered it not in rainforest trees but in the nests of two species of ants that inhabit the Anza-Borrego desert in southern California. The nests of these ants actually do tend to be spaced in a kind of chequerboard. The nearest neighbours of nests of one species, they found, are almost always nests of the other species, just as the nearest neighbours of a black square on a chequerboard are white squares.

The average distance between pairs of nests of the same species is about fifteen metres, and the spacing between them is fairly even. Ryti and Case soon discovered a reason for the regularity. Foragers and soldiers habitually range out from established nests until they meet foragers from other established nests. So long as these other nests are at least fifteen metres away, these interactions are relatively mild, like scouts from two different armies skirmishing occasionally in no man's land. But if a queen of the same species tries to establish a new nest nearer to such an established community, foragers from the older nest will quickly find and destroy it. Ants of the other species do the same thing with newly-established nests of their own species.

Because the foragers from one species completely ignore nests of the other species, the regular chequerboard spacing of the two kinds of nest is simply an accident, due to the fact that both species tend to space their nests at roughly equal distances. If an ecological disturbance, disease, or the death of a queen wipes out one of these nests, a new nest can become established in the gap that has opened up. If this does not happen, the system is remarkably stable.

In the case of the ants, the exclusionary region extending from each nest is caused not by disease organisms but by the foragers and soldiers that are determined to keep the immediate area clear of competing nests of the same species. It is sharply limited in size by the distance the foragers can venture.

When Ryti and Case tried to model these nests in the computer, however, they found that they could not fit more than two or three species into the chequerboard before the density of ant nests became too great. Larger numbers of species would have quickly exhausted the food supply in this sandy, sparsely vegetated desert.

Even though the rainforest is a much richer environment, it seems unlikely at first blush that it could support a complex chequerboard of hundreds of different species, some rare and some common. But perhaps it could if the exclusionary regions surrounding each tree, particularly trees of rare species, were very strong and very extensive in size.

It is possible to imagine a rainforest in which many different species of tree are scattered thinly about, each surrounded by an extensive *cordon sanitaire* that keeps its relatives at great distances. The result would be a jigsaw puzzle of diversity. But this scheme only works if the effects of these exclusion zones are very strong, producing huge and very clearly defined regions in which members of the same species cannot grow.

The need to suppose such strong exclusion zones is rooted in the nature of the Janzen-Connell effect itself. If there really are Janzen-Connell niches in the rainforest, then they form a kind of photographic negative of a standard ecological niche. The pathogens permit all the other tree species in the forest to invade a tree's exclusion zone quite freely. The only trees that are excluded are those that are of the same species, a tiny minority of the total. So the zones will tend to have a rather weak overall effect on rainforest diversity.

None the less, exclusion zones of pathogens and predators really do exist in the rainforest. They have been detected by many surveys and experiments during the quarter-century since the Janzen-Connell hypothesis was proposed. Most of the experiments consist simply of planting clusters of seeds near the parent tree and at varying distances from it, to see how many will survive. Such experiments have been tried using many different species of trees, in rainforests as diverse as those of Costa Rica, Southeast Asia and northern Queensland. The results showed that seeds and seedlings near the parent tree are more likely to die or be eaten than those that are some distance away. The seedlings growing near the base of the parent tree, it turned out, often fall victim to fungus infections, which you will remember is just what had happened to Henry Ford's rubber trees in Fordlandia. So there really are exclusion zones of pathogens around trees in the rainforest, and many different species (though not all) have them. But the effects of these exclusion zones, while statistically significant, are slight, and they usually do not extend more than a few metres from the trunk of the parent tree.

In many cases the zones are not strong enough to overcome completely the natural tendency of seedlings of a tree to be most abundant near the parent tree. Stephen Hubbell showed this using his enormous collection of data from Barro Colorado Island. He found that, contrary to the impressions Alfred Russel Wallace had received as he wandered through the forest, most trees of the same species really do tend to be grouped. The trees are not as tightly clustered as they would be if their seeds were all free to germinate, but they are far more clustered than people had thought. Hubbell's extensive data reinforced the growing consensus among ecologists that exclusion zones, although they exist, tend to be short-range and weak.

On the basis of his unarguable data, Hubbell rejected the Janzen-Connell hypothesis (at least in its pristine form) as the primary explanation for rainforest diversity. At first he was driven, from this and other observations, to a conclusion that I find very discomfiting. He suspected that all this diversity is maintained simply because there are so many trees. Even a rare species of tree has many representatives in the huge forests that still remain on the planet. As a consequence it will take a long time for such a rare species to become extinct, perhaps hundreds of thousands of years, even if there are no forces tending to maintain it. During all this long time evolution has been beavering away among all the other species, so that by the time this species finally does die out at least one new species has appeared somewhere in the forest to take its place.

I must confess to being unhappy with this rather defeatist view of rainforest diversity, and indeed more recently Hubbell himself has moved away from it. The rainforest is filled with frantic, roistering life, with species fiercely competing and continually evolving new ways to attack or foil their competitors. Surely these are the forces that help to push speciation, and that somehow maintain all the diversity once it is produced. If there were no such strong forces, then after a while the forest would be filled with a few great families of closely related trees, each of which could only be divided into species with great difficulty by hair-splitting biologists seizing on whatever differences have happened to accumulate among them by chance.

Species herd-immunity

The Janzen-Connell effect by itself seems inadequate to explain rainforest diversity – it is clear that most of the trees are simply not arranged in some kind of vast chequerboard. But suppose the rainforest has also been shaped by another process? This process would be analogous to the genetic herd-immunity that Doug Green and I had earlier postulated for humans and for other genetically polymorphic animal species.

Herd-immunity operating on species and herd-immunity operating on genes turn out to have remarkably similar properties. Pathogens should act to maintain many different species of tree within a forest just as genetic herd-immunity should maintain many different genotypes within a species. Once a species of tree becomes too common in the rainforest, its pathogens will increase in numbers as well, soon spreading from one tree to another. This drives down the numbers of that species of tree, leaving room for trees of other species to take their place. If some of these species in turn become so numerous that their own pathogens can spread, then they too are driven down in numbers.

How does species herd-immunity differ from the Janzen-Connell effect? In the Janzen-Connell effect, the number of pathogens also increases with the number of host trees, but the pathogens are for the most part closely grouped around each host tree. In species herd-immunity, on the other hand, the spread of the pathogens need not necessarily be a function of how far apart the trees are from each other. Trees of the same species will be free to form limited clusters, so long as these do not become too extensive or too dense.

And pathogens need not be confined to the area immediately around a host tree. Pathogens can often spread rapidly over wide areas by the agency of wind and water and sometimes through less expected routes. Steve Hubbell, working with other ecologists, has recently found that fungal infection can spread from tree to tree through their root systems as long as their roots are touching, even though the trees themselves may be some distance apart.

I attempted to model such a situation, by applying a little gentle species herd-immunity selection to a forest of trees living in the computer. The computer program allowed trees to fall and die here and there in the forest, and each time this happened the program scanned the twenty-five trees that were closest to the downed tree to determine the species that they belonged to. One of the rarer of them was allowed to reproduce and replace the downed tree. This very mild selection maintained many different species in the forest over long periods of time. Fascinatingly, the trees inhabiting the computer tended to form into clusters that built up and then dispersed to be replaced later by other clusters forming elsewhere in the forest. These dynamic fluctuations could continue indefinitely.

Steve Hubbell and I are now looking at his patch of rainforest to see if such a pattern can be detected. Francisco Dallmeier and Jim Comiskey from the Smithsonian Institution have joined the search, using similar data sets that they and others have collected from various tropical rainforests. Hubbell's data are particularly useful for such a search, for it is possible to locate all the trees that have died over a ten-year period and all the young trees that have appeared over the same time. Eighty of the more than three hundred species are common enough for us to perform a statistical analysis on them. We find that in two-thirds of these species there is an excess of births over deaths in regions where the species is rare and an excess of deaths over births in regions where it is common. Further, trees tend to reproduce more readily when they are surrounded by trees of many other kinds – in the rainforest, there really is safety in diversity.

This is just the sort of pattern that the model would predict. There should be enhanced mortality when members of the same species are crowded together, so that their specialized pathogens can spread more easily. And there should be enhanced survival when they are scattered thinly about and protected by all the other species in the forest. Much more needs to be done to rule out other possible explanations, but at first blush it seems that species herd-immunity really is operating in the rainforest.

The rainforest is an ideal place to look for such herd-immunity effects, because there are so many species of tree and so many predators and pathogens. And, of course, there is another advantage as well. The trees stay put until they die, so that the impact of pathogens on the distribution of trees in a rainforest should be far easier to detect than the impact of pathogens on mobile populations of mice or humans. Doug Green and I are sufficiently encouraged by these results to return to the search for traces of genetic herd-immunity in our own species.

The changing role of plagues

Diseases have played, and continue to play, a large role in shaping the diversity of living things on the earth. My guess is that the more we study this process, the more we will discover the true extent of their role. One important result will be a growing realization that in order for us to preserve the diversity of a complex ecosystem like a rainforest or a coral reef, it will be necessary to preserve the pathogens as well as their hosts.

We too have been shaped by our diseases, perhaps to a greater degree than we realize. Much of our biochemical and immunological diversity, as I have suggested, may have been produced by the necessity of keeping many different diseases at bay simultaneously. Is it possible, I wonder, that some of our genetic predisposition towards behavioural diversity might have resulted from this necessity as well? Those of us with a genetic predisposition to be particularly active and exploratory might be confronted with a greater variety of diseases during our lifetimes. Those of us who pass our days as couch potatoes might avoid that peril, only to be faced with others. People who live such sedentary lives are perhaps more likely to be bitten by the insect vectors of disease, and to be exposed for longer periods to the particular pathogens with which they share a single environment – and which have perhaps specialized to prey on couch potatoes.

The violent outbreaks of disease that we call plagues have probably played a rather small part in the evolution of all this diversity. Instead they have, until recently, performed a grimmer role. They have kept the numbers of our species in check, just as they have done repeatedly with most other species of animal and plant on the earth. But now, in spite of the frightening recent outbreaks of plagues and near-plagues that have taken place around the world, their role in controlling the numbers of our own species is fading away.

Plagues have been losing their grip on our own species for longer than most of us realize. As I pointed out in a recent book, *The Runaway Brain*, the evolution of our ancestors was catapulted in a different direction from that taken by most other animals. We have moved, of course, towards dramatically larger brains. At the same time, we have moved away from intimate and dangerous interactions with the natural environment and into an artificial environment of our own making.

All this has been the result of an evolutionary feedback loop. We have become more and more ingenious at inventing ways to divorce ourselves from the perils of the world of nature that surround us. And this enhanced security has provided us with the margin of safety needed to think up even better ways to divorce ourselves from that world.

The evolutionary feedback loop is made up of three interconnected parts. Our increasingly clever brains, of course, make up the first part. The second part consists of our evolving bodies, which have given us – or at least some of us – the capability of carrying out the most complex instructions of our very clever brains. And the third part is our ever more complex and ever more human-oriented environment, which has encouraged the survival of those of us who have the brains and bodies that are best able to take advantage of the opportunities that environment provides.

Until recently, while we were certainly clever enough to have done so, we had not developed the knowledge to understand and overcome our plagues. Yet our ancestors had actually begun to conquer their plagues and other diseases even before they had invented the formal technology. The Danish farmers of the mid-nineteenth century, as they adopted slightly more advanced animal husbandry techniques, were making their environment more human-oriented, and at the same time – as it turned out – safer for them. This small change was enough to break the cycle of the malaria parasite.

Now, at the end of the second millennium, we have instantaneous communications and the ability to transport drugs and chemicals to the remotest parts of the world in days or hours, as was done with the Indian plague. This makes it very unlikely that a plague will decimate our own species before we finally learn to control our numbers by other, more sensible means. Even relatively intractable diseases like AIDS and TB should be in principle controllable. And though abortive plagues may increase in frequency in the future, they will not reach the monstrous numbers of plagues that afflicted China during the sixteenth and seventeenth centuries – at least seventy during that 200-year period. And those, remember, were full-blown plagues, causing hundreds of thousands or even millions of deaths.

But in the meantime our growing numbers are simplifying other ecosystems, causing plagues to spread among far more helpless animals and plants, and forcing them towards extinction. The waterfowl that innocently alight in the polluted tidal pools of northern California, the animals and plants of the rainforest that are exposed to new pathogens as the forest becomes converted to cropland, are only a small part of this sad and growing story.

We are, despite manifold setbacks and stupidities, reducing the toll of plagues on our own species. But we are also, quite accidentally, increasing the toll on others. And this will have a great impact on our long-term survival, and on the overall health of the planet. It will almost certainly be greater than the diminishing impact of diseases on our own species.

GLOSSARY

Anopheles mosquitoes   Mosquitoes of this genus are the chief insect vectors of the human malarias.

Antibody   A protein produced by our immune system that can bind to one or more foreign proteins or other molecules. This binding usually sets in train a number of processes that can destroy the foreign invaders.

Antigen   A protein or other molecule that can stimulate the production of antibodies.

Bases   The order in which the four types of bases (called in biological shorthand A, T, G and C) are arranged along the DNA molecule encodes its genetic information.

Bejel   A yaws-like disease, transmitted primarily by body contact, widespread until recently in North Africa and the Middle East.

Black Death   The most famous of the many outbreaks of bubonic plague. It peaked in Europe in 1348. It did not receive its dread modern name until 1823, being called by many other names before that time.

Buboes   The swollen lymph nodes, particularly in the groin and armpits, characteristic of bubonic plague victims.

Bubonic plague   Caused by the bacillus _Yersinia pestis_. The bubonic form is spread by rat fleas, while the more contagious pneumonic form is spread through the air.

CD4   A molecule found on the surface of helper T-cells and a few other cells in the body. It normally binds to MHC molecules on the surface of other cells and is directly involved in the normal functioning of the immune system. A major protein on the surface of the AIDS viruses is also able to bind to CD4, and this allows the viruses to invade the T-cells.

Cholera	Severe, life-threatening diarrhoea with a sudden onset, caused by *Vibrio cholerae*, a gram-negative comma-shaped bacillus.
Codon	A sequence of three bases in the DNA. Each codon codes for one of the amino acids in a protein.
Cryptosporidium	A protozoan parasite, widespread in the Third World, that causes severe vomiting and diarrhoea.
DNA (deoxyribonucleic acid)	The genetic material of most of the organisms on the planet. These long molecules, in the form of a double helix, carry genetic information coded by a sequence of bases, like letters in a sentence, that form part of their structure.
Ecological niche	The place occupied, and the resources exploited, by a species of animal or plant. It appears that two different species cannot occupy the same ecological niche simultaneously – if they seem to do so, it is probably because of our own ignorance of what constitutes their ecological niches.
Enteric infections	A class of infections, the major killers in the Third World, that result in severe and debilitating vomiting and diarrhoea.
Epidemic	An outbreak of a disease that is usually confined in time and space.
Epizootic	The animal equivalent of a human epidemic.
Escherichia coli (*E. coli*)	A very common bacterial denizen of our guts and those of many other animals. *E.*

coli is usually harmless, but it can cause severe and life-threatening vomiting and diarrhoea and is a large contributor to childhood mortality in the Third World.

Evolutionary trees Organisms, or sequences of genes from organisms, can be arranged in trees, in which the degree of relationship is assumed (not always accurately) to reflect the degree of evolutionary relationship.

Falciparum malaria The severest form of human malaria, which can often invade the brain. Now largely confined to the tropics. Parasites of this species cannot form hypnozoites.

Four Corners Virus A hantavirus that broke out recently in the American Southwest – unlike most of its relatives, it does not cause internal bleeding but rather triggers a severe, life-threatening pneumonia.

Hantaviruses RNA viruses that normally cause severe internal bleeding.

Helicobacter A spiral bacillus that can produce stomach and duodenal ulcers. It is widespread in the Third World, where it can be life-threatening.

Helper T-cells These cells stimulate the production of specific antibodies, and also assist killer T-cells in the destruction of infected cells. These are the cells that are directly invaded and eventually killed by the AIDS viruses.

Herd-immunity If the majority of a group of animals is immunized, they will protect the susceptible minority by preventing the pathogens from spreading and multiplying to high numbers.

HIV (Human Immunodeficiency Virus) An RNA retrovirus. HIV-1 and HIV-2 cause AIDS, though the former is more severe and more infectious than the latter.

Hypnozoites	The dormant form of vivax and other malarias that can survive in the cells of the liver for one or two years. When they emerge from the liver cells, they cause a malaria relapse.
Immune system	A complex set of defences which has evolved to a high degree of efficiency in higher organisms. We possess three major types of immunity: *humoral*, which employs antibodies in our circulation to deal directly with pathogens that have entered our bloodstreams; *mucosal*, which protects us from invasion through the small intestine and the lungs; and *cellular*, which can detect our cells that have been invaded by pathogens and destroy them.
Invasin	A protein made by most *Yersinia* bacteria – but notably not by *Y. pestis* – that allows them to invade the cells lining the guts of their hosts.
Killer T-cells	A class of T-cells that can directly destroy a cell infected by a virus or other pathogen. Killer T-cells are not directly invaded by the AIDS viruses, but they are damaged by HIV infection in ways that are not yet fully understood.
Koch's postulates	Robert Koch's rigorous requirements for proof that a particular pathogenic organism causes a particular disease. It must always be culturable from its victims, be capable of being grown in a pure culture, produce the disease on inoculation into new animals, and be isolable from these animals in turn.
Leprosy	A slowly progressing disease, caused by a relative of the TB bacillus, *Mycobacterium leprae*. Leprosy is difficult to catch, but

once caught it will slowly destroy the peripheral nerves. Because of the resulting numbness, people with leprosy injure themselves repeatedly, and this contributes to the gradual destruction of fingers, toes and other parts of the body.

M (microfold) cells

Cells in the Peyer's patches of the intestine that engulf and usually destroy pathogens.

MHC (Major Histocompatibility Complex)

A set of genes (on chromosome number 6 in humans) that code for a great variety of proteins. Some of these are MHC molecules, which fall into two classes. One class teaches the immune system the difference between self and non-self. The second presents non-self antigens to the helper T-cells, triggering the appropriate immune response.

Natural selection

This process, originally suggested by Darwin, ensures that the genes of the organisms best able to leave offspring survive. Natural selection depends ultimately on mutations, which provide the genetic variation that it sorts out. Note that mutations are arising in all populations all the time, and that only a tiny minority of them will aid their carrier's survival under any particular set of circumstances.

nef

A rapidly evolving HIV gene which, among other things, forces the invaded cells to destroy their own CD4 molecules.

Niche

See ecological niche.

Pandemic

A widespread epidemic that may in some cases spread rapidly around the planet.

Pasteurella multocida

The causative agent of avian cholera, which can also cause a variety of severe diseases in humans and other animals.

Pathogen — Any disease-causing organism.

PCR — The polymerase chain reaction, an extremely clever laboratory technique that allows tiny amounts of a particular piece of DNA to be multiplied millions of times in a test tube.

Peyer's patches — Clumps of cells, on the inner surface of the small intestine, that are the body's chief producers of antibodies.

Plague — An outbreak of a very severe, life-threatening disease. Plagues may be epidemic or pandemic.

Plasmid — A small circular DNA molecule that is often able to pass from one bacterium to another, even across species barriers. Plasmids can carry genes for virulence, and more recently have been found to carry genes for antibiotic resistance.

Plasmodium — Single-celled protozoa that multiply in the red blood cells and cause malaria. There are four malaria parasites that infect humans (*malariae, falciparum, vivax,* and *ovale*) and hundreds of others that cause malaria and allied diseases in a wide variety of animals and birds.

Pneumonic plague — The worst infectious form of bubonic plague, in which the bacteria invade the lungs, destroying the tissue and spreading on the victims' breath.

Polymorphism — The term means 'many forms'. Polymorphism may be phenotypic, affecting the organism's visible characteristics or phenotype – the great ranges of hair and eye colours in the human population is an example. Or it may be genetic, as in the phenotypically invisible but genetically very important ABO blood groups.

Protease   A protein that breaks down other proteins.

Protozoan   A single-celled organism which, like the cells of our own bodies, has a true nucleus and has its genes organized into a number of chromosomes. Protozoa have metabolisms very like ours, which makes them hard to kill.

Quinoline   This antimalarial compound, derived from the bark of the cinchona tree, has been modified in many different ways by chemists in an attempt to keep pace with the appearance of resistant strains of *Plasmodia*.

Rehydration therapy   A mixture of salts and sugars can replace liquids lost through vomiting and diarrhoea, when water or a solution of salts alone will not work. This simple therapy has saved hundreds of thousands of lives.

Retrovirus   An RNA virus that uses the enzyme reverse transcriptase to make DNA copies of its chromosome. These copies can insert themselves into the chromosomes of some of the cells of its host.

Reverse transcriptase   An enzyme that makes a DNA copy of an RNA molecule. This unusual enzyme reverses part of the usual flow of genetic information, which is from DNA to RNA to protein.

RNA (ribonucleic acid)   The genetic material of many viruses. RNA has a structure very similar to that of DNA, although it is usually single-stranded. Like DNA, RNA can carry genetic information.

Salmonella   A large family of bacteria, currently

containing over 1500 different strains, which includes the bacteria that cause typhoid fever and are often responsible for food poisoning.

Shigella A very close relative of *E. coli*, one of the most dangerous and widespread enteric bacteria.

Spirochete see treponeme.

Transmission-blocking immunity Antibodies raised in the human host can be used to destroy the malaria parasites as they multiply in the mosquitoes, blocking the transmission of the parasites through their insect vector.

Transposon A piece of DNA that can move about from chromosome to chromosome, and even from one species to another. These differ from plasmids, which are normally a separate little chromosome, while transposons (rather like the AIDS virus) can insert themselves into a host chromosome, and may not emerge until many generations have passed. Sometimes they never re-emerge but remain a permanent part of the chromosome that they have invaded.

Treponeme A general term for the spiral-shaped bacteria that cause syphilis and a variety of other diseases.

Tuberculosis This widespread disease is caused by the slow-growing bacillus *Mycobacterium tuberculosis*. Most cases of TB are inactive, but occasionally the bacillus overcomes the host's defences and spreads through the body. It can affect many tissues besides the lungs and is very difficult to cure.

Typhoid A severe fever triggered by the bacterium *Salmonella typhi*, often accompanied by tissue destruction and perforation of the intestines.

Typhus
A disease spread by lice, fleas and mites, caused by a tiny bacterium called a rickettsia. Its symptoms bear some resemblance to those of typhoid.

Vibrio cholerae
The comma-shaped bacillus that causes cholera.

Vivax malaria
A relatively mild but very widespread malaria. Vivax malaria can relapse, because it can form little dormant cells called hypnozoites in the liver.

Yaws
A tropical disease which has a wide range of symptoms similar to those of syphilis but is not usually sexually transmitted. Caused by the spirochete *Treponema pallidum* var. *pertenue*.

Yersinia
A family of bacteria that cause bubonic plague and a variety of milder diseases. Most of these species are common in the soil, and some can live in the cells lining our guts.

NOTES

1 The delicate balance between life and death

p. 7 A look at the long history of our manipulation of the environment is in Clive Ponting, *A Green History of the World: the environment and the collapse of great civilizations* (New York: Penguin Books, 1991).

p. 8 The effects of our ancestors on the recent waves of extinction are examined in Paul S. Martin and Richard G. Klein ed., *Quaternary Extinctions: a Prehistoric Revolution* (Phoenix: University of Arizona Press, 1984).

p. 10 Newspaper and magazine accounts, too many to cite here, have chronicled the desperate plight of the Rwandan refugees. The early history of the plague in India is recounted in David Arnold, *Colonizing the Body: State Medicine and Epidemic Disease in Nineteenth Century India* (Berkeley: University of California Press, 1993).

p. 12 This chronology was drawn from local newspaper accounts and interviews with some of the principals involved.

p. 15 The chronicle of how the Four Corners virus was tracked down is in L. E. Chapman and R. F. Khabbaz, 'Etiology and epidemiology of the Four Corners hantavirus outbreak', *Infectious Agents and Disease* 3 (1994): 234–44.

p. 18 A balanced view of these outbreaks can be found in Stephen S. Morse ed., *Emerging Viruses* (New York: Oxford University Press, 1993).

p. 18 The latest figures on the Russian diphtheria outbreak are in J. Maurice, 'Russian chaos breeds diphtheria outbreak', *Science* 267 (1995): 1416–17.

p. 19 The doomsday scenario is explored in Richard Preston, *The Hot Zone* (New York: Random House, 1994) and Laurie Garrett, *The Coming Plague* (New York: Farrar, Straus and Giroux, 1994).

p. 21 See Mirko Grmek, *History of AIDS: emergence and origin of a modern pandemic.* Translated by R. C. Maulitz and J. Duffin (Princeton, NJ: Princeton University Press, 1990).

p. 24 Murphy's analysis is in Philip M. Murphy, 'Molecular mimicry and the generation of host defense protein diversity', *Cell* 72 (1993): 823–6.

2 The penumbra of disease

p. 30 Guerrant's insightful paper is R. L. Guerrant, 'Lessons from diarrheal diseases; demography to molecular pharmacology', *Journal of Infectious Diseases* 169 (1994): 1206–18.

p. 33 The appalling infant mortality in the Peruvian highlands is documented in K. de Meer, R. Bergman, and J. S. Kusner,

'Socio-cultural determinants of child mortality in southern Peru: including some methodological considerations', *Social Science and Medicine* 36 (1993): 317–31.

p. 33 These mortality figures are drawn from the World Health Statistics Annuals.

p. 35 Something about *verruga*, also known as bartonellosis, is in U. Garcia-Caceres and F. U. Garcia, 'Bartonellosis: an immunosuppressive disease and the life of Daniel Alcides Carrión', *American Journal of Clinical Pathology* 95 (suppl.) (1991): S58–S66.

3 The worst of times

p. 37 Graunt's analyses are found in John Graunt, *Reflections on the weekly bills of mortality for the cities of London and Westminster, and the places adjacent* (London: Samuel Speed, 1665) and William Petty, *The economic writings of Sir William Petty, together with the Observations upon the bills of mortality, more probably by Captain John Graunt. Edited by Charles Henry Hull* (New York: A. M. Kelly, 1963–4).

p. 41 The Catal Huyuk data are in James Mellaart, *Catal Huyuk: a neolithic town in Anatolia* (London: Thames & Hudson, 1967).

p. 42 The data in these graphs were drawn from John Marshall, *Mortality of the metropolis* (London: J. Haddon, 1832).

p. 44 Ewald's book is Paul W. Ewald, *Evolution of Infectious Disease* (New York: Oxford University Press, 1994).

p. 51 The title of part two is from the poet William Cowper's *Table Talk*.

4 Four tales from the New Decameron

p. 53 The story of the muskoxen is recounted in J. E. Blake, B. D. McLean, and A. Gunn, 'Yersiniosis in free-ranging muskoxen in Banks Island, Northwest Territories, Canada', *Journal of Wildlife Diseases* 27 (1991): 527–33.

p. 55 Procopius' account is in Procopius, *History of the Wars, Books I and II* ed. H. B. Dewing, Vol. I (Cambridge, Mass.: Harvard University Press, 1914). Some of the demographic consequences of the Plague of Justinian are discussed in J. C. Russell, 'That earlier plague' *Demography* 5 (1968): 174–84.

p. 59 An account of early Chinese plagues, and a chronology of disease outbreaks, is found in Lien-Teh Wu, J. W H. Chun, R. Pollitzer, *e al.*, *Plague: a Manual for Medical and Public Health Workers* (Shanghai: National Quarantine Service, 1936).

p. 60 Michael W. Dols, *The Black Death in the Middle East* (Princeton, NJ: Princeton University Press, 1977) gives an account of the history of the plague at the world's crossroads.

p. 60 Discussions of the origin of the Black Death appear in William H. McNeill, *Plagues and Peoples* (Garden City, NY: Anchor Press, 1976) and J. Norris, 'East or west? The geographic origin of the Black Death',

Bulletin of the History of Medicine 51 (1977): 1–24. Two useful books out of hundreds about the Black Death are Michael W. Dols, *The Black Death in the Middle East* (Princeton, NJ: Princeton University Press, 1977) and Robert S. Gottfried, *The Black Death: natural and human disaster in medieval Europe* (New York: Free Press, 1983).

p. 65 Tales of rats in present-day New York are found in: S. Erlanger, 'In New York, rats survive the man race', *New York Times*, 1987, New York, pp.B1, B2. Horror stories about the superb adaptability of rats in general abound in Robert Hendrickson, *More Cunning Than Man* (New York: Stein and Day, 1983).

p. 66 Graham Twigg, *The Black Death: a biological reappraisal* (New York: Schocken Books, 1984) is a revisionist view of the Black Death with which few will agree, but it has an excellent discussion of the role of rats and other animals.

p. 69 An account of Eyam's terrible ordeal can be found in Walter George Bell, *The Great Plague in London in 1665* (London: John Lane, Bodley Head, 1924).

p. 69 Slack's ideas are in Paul Slack, *The Impact of Plague in Tudor and Stuart England* (London: Routledge, 1985).

p. 72 An excellent account of Alexandre Yersin's life can be found in E. Bendiner, 'Alexandre Yersin: pursuer of plague', *Hospital Practice* 24 (1989): 121–38. See also H. H. Mollaret & Jacqueline Brossollet, *Yersin, un pasteurien*

en Indochine (Paris: Belin, 1993).

p. 77 The discovery of invasin is reported in R. R. Isberg and S. Falkow, 'A single genetic locus encoded by *Yersinia pseudotuberculosis* permits invasion of cultured animal cells by *Escherichia coli* K-12', *Nature* 317 (1985): 262–4.

p. 81 The properties missing in *Y. pestis* are listed in R. R. Brubaker, 'Molecular biology of the dread Black Death', *ASM News* 50 (1984): 240–5. D. J. Sikkema and R. R. Brubaker, 'Resistance to pesticin, storage of iron, and invasion of HeLa cells by Yersiniae', *Infection and Immunity* 55 (1987): 572–8 describes the inability of *Y. pestis* to invade human cells.

p. 82 Wolf-Watz's research on the virulence of *Yersinia* is in R. Rosqvist, M. Skurnik, and H. Wolf-Watz, 'Increased virulence of *Yersinia pseudotuberculosis* by two independent mutations', *Nature* 334 (1988): 522–5.

p. 83 The work of Quan and his group is in O. A. Sodeinde, Y. Subrahmanyam, K. Stark, T. Quan, Y. Bao and J. D. Goguen, 'A surface protease and the invasive character of plague', *Science* 258 (1994):1004–7.

p. 84 The relationship between *Yersinia* and other pathogens is recounted in M. Krause, J. Fierer and D. Guiney, 'Homologous DNA sequences on the virulence plasmids of pathogenic *Yersinia* and *Salmonella dublin* Lane', *Molecular Microbiology* 4 (1990): 905–11 and S. Håkansson, T. Bergman, J-C. Vanooteghem, *et al.*, 'YopB and YopD constitute a novel class of *Yersinia* Yop proteins', *Infection and Immunity* 61 (1993): 71–80.

p. 85 E. S. Williams, E. T. Thorne, T. J. Quan, *et al.*, 'Experimental infection of domestic ferrets and Siberian polecats with *Yersinia pestis*', *Journal of Wildlife Diseases* 27 (1991): 441–5 shows how plague bacilli might persist in other animals.

p. 85 The flea story is in K. A. McDonough, A. M. Barnes, T. J. Quan, *et al.*, 'Mutation in the *pla* gene of *Yersinia pestis* alters the course of the plague bacillus-flea interaction'. *Journal of Medical Entomology* 30 (1993): 772–80. An account of the *Yersinia* proteases is in E. E. Galyov, S. Håkansson, A. Forsberg, *et al.*, 'A secreted protein kinase of *Yersinia pseudotuberculosis* is an indispensable virulence determinant', *Nature* 361 (1993): 730–2.

5 Was the Indian plague actually plague, and if not why not?

p. 91 Some of the day-to-day events and rumours about the Indian plague were gleaned from stories in the *Times of India* and other Indian newspapers for late September 1994, along with discussions with many people throughout India. The *New York Times* also reported some of the confusion: L. K. Altman, 'Was there or wasn't there a pneumonic plague epidemic?', *New York Times*, Nov. 15 1994, New York, pp. C1, C3.

p. 92 The dissenting letters from Indian scientists were: L. Dar, R. Thakur and V. S. Dar, 'India: is it plague?', *Lancet* 344 (1994): 1359 and T. J. John, 'India: is it plague?', *Lancet* 344 (1994): 1359–60.

p. 93 Some of the conditions of Calcutta are recounted in E. Eide, 'The coolies of Calcutta', *World Press Review* 40 (1993): 51–3 and Dominique Lapierre, *City of Joy* (New York: Doubleday, 1985).

p. 98 Fascinating stories about Haffkine can be found in T. Lowy, From guinea pigs to man: the development of Haffkine's anticholera vaccine', *Journal of the History of Medicine and Allied Sciences* 47 (1992): 270–309 and E. Chernin, 'Ross defends Haffkine: the aftermath of the vaccine-associated Mulkowal disaster of 1902', *Journal of the History of Medicine and Allied Sciences* 46 (1991): 201–18. The story of the early misadventures with the Salk vaccine is told in Jane S. Smith, *Patenting the Sun: polio and the Salk vaccine* (New York: William Morrow, 1990).

6 Cholera, the Black One

p. 105 A survey of the history of the cholera pandemics is found in P. Shears, 'Cholera', *Annals of Tropical Medicine and Parasitology* 88 (1994): 109–22.

p. 106 Mary Jesudason's discovery of cholera strain O139 is recounted in M. V. Jesudason, M. K. Lalitha and G. Koshi 'Non O1 *Vibrio cholerae* in intestinal and extraintestinal infections in Vellore, S. India', *Indian*

Journal of Pathology and Microbiology 34 (1991): 26–9 and T. J. John and M. V. Jesudason, 'The spread of *Vibrio cholerae* 0139 in India', *Journal of Infectious Diseases* 171 (1995): 759–60. The possibility that it constitutes an eighth cholera pandemic is investigated in D. L. Swerdlow and A. A. Ries, '*Vibrio cholerae* non-01 – the eighth pandemic?', *Lancet* 342 (1993): 382–3.

p. 106 The spread of the seventh cholera epidemic to Peru was traced at the molecular level by I. K. Wachsmuth, G. M. Evins, P. I. Fields, *et al.*, 'The molecular epidemiology of cholera in Latin America', *Journal of Infectious Diseases* 167 (1993): 621–6.

p. 107 The role of water supply cholera in Trujillo was followed in detail in D. L. Swerdlow, E. D. Mintz, M. Rodriguez, *et al.*, 'Waterborne transmission of epidemic cholera in Trujillo, Peru: lessons for a continent at risk', *Lancet* 340 (1992): 28–33.

p. 108 The contrast between African and Peruvian cholera is discussed in R. I. Glass, M. Claeson, P. A. Blake, *et al.*, 'Cholera in Africa: lessons on transmission and control for Latin America', *Lancet* 338 (1991): 791–5.

p. 110 The Milwaukee outbreak is recorded in W. R. MacKenzie, N. J. Hoxie, M. E. Proctor, *et al.*, 'A massive outbreak in Milwaukee of cryptosporidium infection transmitted through the public water supply', *New England Journal of Medicine* 331 (1994): 161–7.

p. 110 Some of Snow's accounts of his researches can be found in John Snow, *Snow on Cholera* [facsimile of 1936 edition] (New York: Haffner, 1965).

p. 116 Much information on cholera is gathered in D. Barua and W. B. Greenough ed., *Cholera* (New York: Plenum Medical Book Co., 1992). Involvement of nerve cells in cholera toxicity is detailed in M. Jodal, 'Neuronal influence on intestinal transport', *Journal of Internal Medicine. Supplement* 732 (1990): 125–32. The first report of cholera toxin is S. N. De, 'Enterotoxicity of bacteria-free culture-filtrate of *Vibrio cholerae*', *Nature* 183 (1959): 1533–4.

p. 119 Some of Pettenkofer's strange history is in N. Howard-Jones, 'Gelsenkirchen typhoid epidemic of 1901, Robert Koch, and the dead hand of Max von Pettenkofer', *British Medical Journal* 1 (1973): 103–5.

p. 121 The section heading is from John Milton's *Lycidas*.

p. 121 J. Grimes, R. W. Atwell, P. R. Brayton, *et al.*, 'The fate of enteric pathogenic bacteria in estuarine and marine environments', *Microbiological Sciences* 3 (1986): 324–9 shows that *Vibrio* is not alone in being able to enter a dormant state in the marine environment.

p. 122 Some details of the remarkable ecology of the cholera bacillus can be found in M. L. Tamplin, A. L. Gauzens, A. Hug, *et al.*, 'Attachment of *Vibrio cholerae* serogroup 01 to zooplankton and phytoplankton of Bangladesh waters', *Applied and Environmental Microbiology* 56 (1990): 1977–80.

p. 126 The origin of the inserted sequence carrying the toxin genes is investigated by T. Yamamoto, T. Gojobori, and T. Yokota, 'Evolutionary origin of pathogenic determinants in enterotoxigenic *Escherichia coli* and *Vibrio cholerae* O1', *Journal of Bacteriology* 169 (1987): 1352–7.

p. 126 Mekalanos's finding is reported in G. D. N. Pearson, A. Woods, S. L. Chiang, *et al.*, 'CTX genetic element encodes a site-specific recombination system and an intestinal colonization factor', *Proceedings of the National Academy of Sciences (U.S.)* 90 (1993): 3750–4. The truly complex nature of *Vibrio* virulence, and the astonishing number of genes involved, is summarized in V. DiRita, C. Parsot, G. Jander, *et al.*, 'Regulatory cascade controls virulence in *Vibrio cholerae*', *Proceedings of the National Academy of Sciences (U.S.)* 88 (1991): 5403–7.

p. 128 Details of the avian cholera outbreaks are in R. G. Botzler, 'Epizootiology of avian cholera in wildfowl', *Journal of Wildlife Diseases* 27 (1991): 367–95 and S. M. Combs and R. G. Botzler, 'Correlations of daily activity with avian cholera mortality among wildfowl', *Journal of Wildlife Diseases* 27 (1991): 543–50.

7 *A cleverer pathogen*

p. 132 My grandmother's operation is recounted in C. C. Holman, 'Survival after removal of twenty feet of intestine', *Lancet* 2 (1944): 597.

p. 134 The slow decline in typhoid in the last part of the nineteenth century is documented in Robert Woods and John Woodward ed., *Urban disease and mortality in nineteenth century England* New York: St Martin's Press, 1984).

p. 135 Various accounts of Typhoid Mary are given in Major George A. Soper, *Typhoid Mary*, reprinted in *The Carrier State* (New York: Arno Press, 1977) and Anon., '"Typhoid Mary" has reappeared' in *New York Times Magazine* Apr. 4 1915, New York, pp. 3–4. I have drawn on accounts in the *New York Times* of 1 July 1909, 21 Feb. 1910, 29 Nov. 1914, 28 and 31 March 1915 Sara Josephine Baker's account of Typhoid Mary's capture is told in Sara Josephine Baker, *Fighting for Life* (New York: Macmillan, 1939).

p. 138 The involvement of gallstones in the carrier state is examined in C. W. Lai, R. C. Chan, A. F. Cheng, *et al.*, 'Common bile duct stones: a cause of chronic salmonellosis', *American Journal of Gastroenterology* 87 (1992): 1198–9.

p. 141 Huckstep's story is told in Robert L. Huckstep, *Typhoid Fever and Other Infections* (Edinburgh: Livingstone, 1962).

p. 143 E. A. Groisman and H. Ochman, 'Cognate gene clusters govern invasion of host epithelial cells by *Salmonella typhimurium* and *Shigella fexneri*', *EMBO Journal* 12 (1993): 3779–87 details Groisman's discovery.

8 An ague very violent

p. 150 Some of the devastating impact of malaria is recounted in Anon., 'World malaria situation 1990', *World Health Statistics Quarterly* 45 (1992): 257–66.

p. 150 An excellent source of information on malaria is L. J. Bruce-Chwatt, *Essential Malariology* 2nd edn (New York: Wiley, 1985).

p. 160 Ross's own account of his discovery, including much that reveals his character, is found in Ronald Ross, *Memoirs, with a full account of the great malaria problem and its solution* (London: John Murray, 1923).

p. 163 The possible re-emergence of yellow fever is discussed in J. Maurice, 'The fall and rise of yellow fever', *World Press Review* 41 (1994): 48–9.

p. 163 The relative ease with which malaria was conquered in Europe is followed in Leonard Bruce-Chwatt and Julian de Zulueta, *The Rise and Fall of Malaria in Europe: a historico-epidemiological study* (New York: Oxford University Press, 1980).

p. 164 Slater's discovery of the mechanism of quinine derivatives is recounted in A. F. Slater and A. Cerami, 'Inhibition by chloroquine of a novel haem polymerase enzyme activity in malaria trophozoites', *Nature* 355 (1992): 167–9.

p. 167 The story of the Danish mosquitoes is detailed in C. Wesenberg-Lund, *Contributions to the biology of the Danish* Culicidae (Copenhagen: A. F. Holst and Son, 1921).

p. 170 Accounts of Shortt and Garnham's work appeared in H. E. Shortt and P. C. C. Garnham, 'Pre-erythrocytic stages in mammalian malaria parasites', *Nature* (1948): 126 and H. E. Shortt, P. C. C. Garnham, G. Covell, *et al.*, 'The pre-erythrocytic state of human malaria, *Plasmodium vivax*', *British Medical Journal* 1 (1948): 547.

p. 172 Krotoski's account of his discovery is in W. A. Krotoski, 'The hypnozoite and malarial relapse', *Prog. Clin. Parasitol.* 1 (1989): 1–19.

p. 174 The relationship among the malaria parasites is examined in A. P. Waters, D. G. Higgins, and T. F. McCutchan, '*Plasmodium falciparum* appears to have arisen as a result of lateral transfer between avian and human hosts', *Proceedings of the National Academy of Sciences (U.S.)* 88 (1991): 3140–4 and A. P. Waters, D. G. Higgins, and T. F. McCutchan, 'Evolutionary relatedness of some primate models of Plasmodium', *Molecular Biology and Evolution* 10 (1993): 914–23.

p. 176 A recent account of the qualified success of Patarroyo's vaccine is found in J. Maurice, 'Malaria vaccine raises a dilemma', *Science* 267 (1995): 320–3.

p. 177 Huff's early work is summarized in C. G. Huff, 'Recent experimental research on avian malaria', *Advances in Parasitology* 6 (1968): 293–311.

p. 178 Carter's suggestion of transmission-blocking

immunity is explained in R. W. Gwadz, R. Carter, and I. Green, 'Gamete vaccines and transmission-blocking immunity in malaria', *Bulletin of the World Health Organisation* 57 (1979): 175–180.

p. 180 The story of attempts at malaria eradication in Sri Lanka is told in J. Pinikahana and R. A. Dixon, 'Trends in malaria morbidity and mortality in Sri Lanka', *Indian Journal of Malariology* 30 (1993): 51–5.

p. 181 The effect of poverty on malaria incidence is shown in A. C. Gamage-Mendis, R. Carter, C. Mendis, *et al.*, 'Clustering of malaria infections within an endemic population: risk of malaria associated with the type of housing construction', *American Journal of Tropical Medicine and Hygiene* 45 (1991): 77–85.

p. 182 The Taiwan eradication story is in W. I. Chen, 'Malaria eradication in Taiwan'. *Kao-Hsiung i Hsueh Ko Hsueh Tsa Chih Kaosiung Journal of Medical Science* 7 (1991): 263–70.

p. 183 Carter's evidence is in R. Carter and K. N. Mendis, 'Immune responses against sexual stages of *Plasmodium vivax* during human malarial infections in Sri Lanka', *Parasitologia* 33 (1991) 67–70. The difficulties *P. falciparum* encounters in transmission were noted in A. C. Gamage-Mendis, J. Rajakaruna, S. Weerasinghe, *et al.*, 'Infectivity of *Plasmodium vivax* and *P. falciparum* to *Anopheles tessellatus*; relationship between oocyst and sporozoite

development', *Transactions of the Royal Society of Tropical Medicine and Hygiene* 87 (1993): 3–6. Some of the many complexities of malaria transmission are set forth in O. A. Erunkulu, A. V. Hill, D. P. Kwiatkowski, *et al.*, 'Severe malaria in Gambian children is not due to lack of previous exposure to malaria', *Clinical and Experimental Immunology* 89 1992) 296–300 and S. W. Lindsay, J. H. Adiamah, J. E. Miller, *et al.*, 'Variation in attractiveness of human subjects to malaria mosquitoes in The Gambia', *Journal of Medical Entomology* 30 (1993): 368–73.

p. 184 Effect of village size on malaria mortality is measured in A. De Francisco, A. J Hall, J. R. Schellenberg, *et al.*, 'The pattern of infant and childhood mortality in Upper River Division, The Gambia', *Annals of Tropical Paediatrics* 13 (1993): 45–52. The lessened impact of malaria in Kinshasha was examined in J. Coene, 'Malaria in urban and rural Kinshasha: the entomological input', *Medical and Veterinary Entomology* 7 (993): 127–37.

9 Syphilis and the Faustian bargain

p. 185 A good introduction to these events in Columbus's life remains Samuel Eliot Morison, *Admiral of the Ocean Sea, a life of Christopher Columbus* (Boston: Little, Brown, 1942).

p. 188 The story of the Tuskegee study is recounted in James H.

Jones, *Bad Blood: the Tuskegee syphilis experiment* (New York: Free Press, 1981).

p. 193 An account of Fracastoro's poem and the history of the term syphilis is in F. S. Glickman, 'Syphilus', *Journal of the American Academy of Dermatology* 12 (1985): 593–6. A thorough discussion of the confusing book by Diaz de Isla is found in Richmond Holcomb, *Who Gave the World Syphilis?* (New York: Froben Press, 1937).

p. 193 A survey of the history of leprosy is included in K. Manchester and C. Roberts, 'The paleopathology of leprosy in Britain: a review', *World Archaeology* 21 (1989): 265–72. F. Lee and J. Magilton, 'The cemetery of the hospital of St James and St Mary Magdalene, Chichester – a case study', *World Archaeology* 21 (1989): 273–82 examines the kinds of lesions found in these leprosy hospitals.

p. 194 The paper that gained such a consensus for the Columbian theory is B. J. Baker and G. J. Armelagos, 'The origin and antiquity of syphilis', *Current Anthropology* 29 (1988): 703–37.

p. 196 B. M. Rothschild and W. Turnbull, 'Treponemal infection in a Pleistocene bear', *Nature* 329 (1987): 61–2 details the rather unlikely bear story.

p. 198 The genetics and physiology of the feeble syphilis spirochete are examined in I. Saint Girons, S. J. Norris, U. Göbel, *et al.*, 'Genome structure of spirochetes', *Research in Microbiology* 143 (1992): 615–21 and P. Hindersson, D.

Thomas, L. Stamm, *et al.*, 'Interaction of spirochetes with the host', *Research in Microbiology* 143 (1992): 629–39.

p. 200 Hudson's accounts of bejel are found in Ellis H. Hudson, *Treponematosis* (New York: Oxford University Press, 1946) and Ellis H. Hudson, *Non-Venereal Syphilis* (Edinburgh: E. & H. Livingston, 1958).

p. 203 Thomas B. Turner and David H. Hollander, *Biology of the Treponematoses* (Geneva: World Health Organization, 1957) shows the inability to distinguish the treponemes.

p. 204 The role of Pygmies in yaws infections is given in V. Herve, E. Kassa Kelembho, P. Normand, *et al.*, 'Resurgence of yaws in Central African Republic: role of the Pygmy population as a reservoir of the virus' (in French), *Bulletin de la Société de Pathologie Exotique* 85 (1992): 342–6.

p. 205 The difficulty of distinguishing these spirochetes genetically is recounted in G. T. Noordhoek, B. Wieles, J. J. van der Sluis, *et al.*, 'Polymerase chain reaction and synthetic DNA probes: a means of distinguishing the causative agents of syphilis and yaws?', *Infection and Immunity* 58 (1990): 2011–13.

p. 206 Hackett's idea that pinta is the oldest of these diseases is in C. J. Hackett, 'On the origin of the human treponematoses', *Bulletin of the World Health Organization* 29 (1963): 7–41.

p. 207 H. J. Engelkens, J. Judanarso, J. J. van der Sluis, *et al.*, 'Disseminated early yaws:

report of a child with a remarkable genital lesion mimicking venereal syphilis', *Pediatric Dermatology* 7 (1990): 60–2 reports a case of syphilis-like yaws. The emergence of milder forms of yaws is suggested in F. A. Vorst, 'Clinical diagnosis and changing manifestations of treponemal infection', *Reviews of Infectious Diseases* 7, Supp. 2 (1985): S327–31, although this study has been criticized.

p. 210 Riviere's discovery is recounted in G. R. Riviere, M. A. Wagoner, S. A. Baker-Zander, et al., 'Identification of spirochetes related to *Treponema pallidum* in necrotizing ulcerative gingivitis and chronic periodontitis', *New England Journal of Medicine* 325 (1991): 539–43.

10 *AIDS and the future of plagues*

p. 215 Uganda's grim story is told in Robert Caputo, 'Uganda: land beyond sorrow', *National Geographic* 173 (4) (1988): 468–91.

p. 216 The worldwide economic and social impact of AIDS has been examined in many publications. A particularly wide-ranging examination is R. F Black, S. Collins and D. L. Boroughs, 'The hidden cost of AIDS', *U.S. News and World Report* 113 (1992): 48–51. The demographic predictions of Anderson and his co-workers are in R. M. Anderson, R. M. May, M. C. Boily, et al., 'The spread of HIV-1 in Africa: sexual contact patterns and the

predicted demographic impact of AIDS', *Nature* 352 (1991): 581–9. Some of the untoward consequences of predicting the course of AIDS are listed in Anon., 'African apocalypse', *Time* 140 (1992): 21.

p. 218 The decline in world assistance to Africa is chronicled in H. W. French, 'African democracies worry aid will dry up', *New York Times*, Mar. 19 1995, New York, pp. A1, A8.

p. 220 The ghastly history of slavery and forced labour in Africa is a good deal worse than I have had time to paint it here. A few glimpses are provided by A. Zegeye and S. Ishemo ed., *Forced Labour and Migration: patterns of movement within Africa* (London: H. Zell, 1989), P. C. W. Gutkind, R. Cohen, and J. Copans ed., *African Labor History* (Beverly Hills: Sage Publications, 1978), A. T. Nzula, I. I. Potehkin and A. Z. Zusmaovich, *Forced Labour in Colonial Africa* (tr. Hugh Jenkins) (London: Zed Press, 1979) and Bill Freund, *The African Worker* (New York: Cambridge University Press, 1988). A correlation of population upheaval and AIDS incidence is discussed in C. M. Becker, 'The demo-economic impact of the AIDS pandemic in Sub-Saharan Africa', *World Development* 18 (1990): 1599–1620 and C. W. Hunt, 'Migrant labor and sexually transmitted disease: AIDS in Africa', *Journal of Health and Social Behavior* 30 (1989): 353–73.

p. 222 Good summaries of the course of HIV infection and the spread of the epidemic are found in G. Pantaleo, C. Graziosi, and A. S. Fauci, 'The immunopathogenesis of human immunodeficiency virus infection', *New England Journal of Medicine* 328 (1993): 327–35 and S. M. Schnittman and A. S. Fauci, 'Human immunodeficiency virus and acquired immunodeficiency syndrome: an update', *Advances in Internal Medicine* 39 (1994): 305–55.

p. 225 A recent balanced appraisal of Duesberg and his critics is found in J. Cohen, 'The Duesberg phenomenon', *Science* 266 (1994): 1642–4.

p. 228 The story of Pasteur's notebooks is in C. Anderson, 'Pasteur notebooks reveal deception', *Science* 259 (1993): 1117.

p. 229 M. Essex *et al.* ed., *AIDS in Africa* (New York: Raven Press, 1994) provides an excellent and up-to-date survey of the spread of the disease and its various manifestations in Africa.

p. 231 The effects of destroying helper and killer T-cells in mice are followed in C. H. Ladel, I. E. Flesch, J. Arnoldi, *et al.*, 'Studies with MHC-deficient knock-out mice reveal impact of both MHC I- and MHC II-dependent T-cell responses on *Listeria monocytogenes* infection', *Journal of Immunology* 153 (1994): 3116–22.

p. 231 The symptomless group of Australians infected for years by HIV-1 is reported in J. Learmont, B. Tindall, L. Evans, *et al.*, 'Long-term symptomless HIV-1 infection in recipients of blood products

from a single donor', *Lancet* 340 (1992): 863–7. The group of apparently resistant Nairobi prostitutes was widely discussed in the press, but details have yet to appear in the literature. Francis Plummer confirmed their existence to me in conversation.

p. 232 Still an excellent summary of the problems facing AIDS researchers is the special issue of *Science* (28 May 1993) largely devoted to the subject.

p. 234 The central role of *nef* in the virulence of HIV is shown by H. W. Kestler 3rd, D. J. Ringler, K. Mori, *et al.*, 'Importance of the nef gene for maintenance of high virus loads and for development of AIDS', *Cell* 65 (1991): 651–62. The dynamics of CD4 and the HIV virus's effects on it are explored in A. Pelchen-Matthews, J. E. Armes, G. Griffiths, *et al.*, 'Differential endocytosis of CD4 in lymphocytic and nonlymphocytic cells', *Journal of Experimental Medicine* 173 (1991): 575–87 and C. Aiken, J. Konner, N. R. Landau, *et al.*, 'Nef induces CD4 endocytosis: requirement for a critical dileucine motif in the membrane-proximal CD4 cytoplasmic domain', *Cell* 76 (1994): 853–64.

p. 235 The dynamics of HIV-2 infection are followed in P. J. Kanki, K. U. Travers, S. Mboup, *et al.*, 'Slower heterosexual spread of HIV-2 than HIV-1', *Lancet* 343 (1994): 943–6 and R. Marlink, P. Kanki, I. Thior, *et al.*, 'Reduced rate of disease development after HIV-2 infection as compared to HIV-1', *Science* 265 (1994): 1587–90.

p. 234 Y. Li, Y. M. Naidu, M. D. Daniel, *et al.*, 'Extensive genetic variability of simian immunodeficiency virus from African green monkeys', *Journal of Virology* 63 (1989): 1800–2 measures the prevalence of SIVs in African green monkeys. S. W. Barnett, K. K. Murthy, B. G. Herndier, *et al.*, 'An AIDS-like condition induced in baboons by HIV-2', *Science* 266 (1994): 642–6 deals with baboons and HIV-2.

p. 240 The concept of a cloud of different viral mutants is set out in M. Eigen, 'Viral quasispecies', *Scientific American* 269(1) (1993): 42–9.

p. 241 This statistical technique is detailed in M. Nei and T. Gojobori, 'Simple methods for estimating the numbers of synonymous and nonsynonymous nucleotide substitutions', *Molecular Biology and Evolution* 3 (1986): 418–26 and A. L. Hughes and M. Nei, 'Patterns of nucleotide substitution at major histocompatibility complex class I loci reveals overdominant selection', *Nature* 335 (1988): 167–70.

p. 000 The results of my analyses are in C. Wills, A. F. Slater and G. Myers, 'Biphasic evolution of the HIV and SIV viruses' (1995): in preparation.

p. 249 A good summary of the growing problem of TB is given in Anon., 'The global challenge of tuberculosis', *Lancet* 344 (1994): 277–9, and an in-depth study is the excellent book by Frank Ryan, *The Forgotten Plague* (Boston: Little, Brown, 1993).

p. 249 The effectiveness of BCG in preventing leprosy is reported in D. M. Baker, J. S. Nguyen-Van-Tam and S. J. Smith, 'Protective efficacy of BCG vaccine against leprosy in southern Malawi', *Epidemiology and Infection* 111 (1993): 21–5. Mucosal immunity to TB in guinea pigs was enhanced by aerosol BCG and reported in M. Gheorghiu, 'BCG-induced mucosal immune responses', *International Journal of Immunopharmacology* 16 (1994): 435–44.

p. 251 Figures for TB incidence among the homeless of San Francisco are given in A. R. Zolopa, J. A. Hahn, R. Gorter, *et al.*, 'HIV and tuberculosis infection in San Francisco's homeless adults', *Journal of the American Medical Association* 272 (1994): 455–61. The huge numbers of different drug-resistant strains that can be detected are examined in Y. Zhang and D. Young, 'Strain variation in the Kat G region of *Mycobacterium tuberculosis*', *Molecular Biology* 14 (1994): 301–8.

11 Safety in diversity

p. 257 An excellent description of MHC is given in V. H. Engelhard, 'How cells process antigens', *Scientific American* 271 (1994): 54–61.

p. 257 The association of malaria with MHC that later vanished is recounted in A. V. S. Hill, C. Allsopp, D. Kwiatkowski, *et al.*, 'Common West African HLA antigens are associated with protection from severe malaria', *Nature* 352 (1991): 595–600.

p. 261 Accounts of our devious journey to genetic herd-immunity are in C. Wills, 'The maintenance of multiallelic polymorphism at the MHC region', *Immunological Reviews* 124 (1991): 165–220 and C. Wills and D. R. Green, 'A genetic herd immunity model for the maintenance of MHC polymorphism', *Immunological Reviews* 143 (1994): 263–92.

p. 261 The balance between these two forces is discussed in M. A. Nowak, K. Tarczy-Hornoch and J. M. Austyn, 'The optimal number of major histocompatibility complex molecules in an individual', *Proceedings of the National Academy of Sciences (U.S.)* 89 (1992): 10896–9.

p. 262 A very readable account of Marshall's discovery is in T. Monmaney, 'Marshall's hunch', *New Yorker* 69 (1993): 64–72.

p. 262 Some of the multitude of effects of *H. pylori* are recounted in J. R. Glynn, '*Helicobacter pylori* and the heart', *Lancet* 344 (1994): 146, and G. C. Cook, 'Gastroenterological emergencies in the tropics', *Baillieres Clinical Gastroenterology* 5 (1991): 861–86.

p. 264 Some of the complexities of the evolution of MHC are traced in the numerous papers in volume 143 of *Immunological Reviews* (1995).

p. 270 The possibility that plagues actually have a small role to

play in MHC evolution is suggested in J. Klein and C. O'hUigin, 'MHC polymorphism and parasites', *Philosophical Transactions of the Royal Society B* 346 (1994): 351–7.

p. 270 Francis Black's different view is F. L. Black, 'Why did they die?', *Science* 258 (1992): 1739–40.

12 Why so many different diseases?

p. 272 Gillett's seminal paper is J. B. Gillett, 'Pest pressure, an underestimated factor in evolution' in *Taxonomy and Geography, a Symposium*, ed. D. Nichols (London: London Systematics Association, 1962). The original papers detailing the Janzen-Connell hypothesis are J. H. Connell, 'On the role of natural enemies in preventing competitive exclusion in some marine animals and in rain forest trees' in *Dynamics of Populations* ed. P. J. Den Boer and G. Gradwell (New York: PUDOC, 1971) and D. H. Janzen, 'Herbivores and the number of tree species in tropical forests', *American Naturalist* 104 (1970): 501–29.

p. 273 Some books about the dynamics of the neotropical rainforest and about Manù in particular are: Kim MacQuarrie, *Peru's Amazonian Eden: Manù National Park and Biosphere Reserve* (Francis O. Patthey and Sons, 1986), D. H. Janzen ed., *Costa Rican Natural History* (Chicago: Chicago University Press, 1983), Adrian

Forsyth and Kenneth Miyata, *Tropical Nature* (New York: Scribner's, 1984) and John Terborgh, *Diversity and the Tropical Rain Forest* (New York: W.H. Freeman, 1992).

p. 279 R. Bierregard, T. Lovejoy, V. Kapos, *et al.*, 'The biological dynamics of tropical rainforest fragments: a prospective comparison of fragments and continuous forest', *BioScience* 42 (1992): 859–67 examines the effect of habitat fragmentation.

p. 280 S. P. Hubbell, R. P. Foster, R. Condit, *et al.*, 'Light gaps and tree diversity in a neotropical forest' (1995): submitted, shows how the gaps in the forest left by fallen trees are filled in.

p. 282 The story of Amazonian rubber and Ford's failed attempt can be found in Richard Collier, *The River that God Forgot; the Story of the Amazon Rubber Boom* (New York: Dutton, 1968).

p. 287 Hubbell's demonstration of clustering of trees in the rainforest is in S. P. Hubbell, 'Tree dispersion, abundance and diversity in a tropical dry

forest', *Science* 203 (1979): 1299–1309.

p. 285 The ecological chequerboard of ant species is found in R. T. Reti and T. J. Case, 'The role of neighborhood competition in the spacing and diversity of ant communities', *American Naturalist* 139 (1992): 355–74.

p. 287 Some of the conflicting experimental results on the Janzen-Connell effect are in C. K. Augspurger, 'Seed dispersal of the tropical tree *Platypodium elegans*, and the escape of its seedlings from fungal pathogens', *Journal of Ecology* 71 (1983): 759–71.

p. 287 The dynamics of fungi spreading beneath the ground in the rainforest are followed in G. S. Gilbert, S. P. Hubbell and R. B. Foster, 'Density and distance-to-adult effects of a canker disease of trees in a moist tropical forest', *Oecologia* 93 (1994): 100–8.

p. 291 The evolutionary feedback loop that led to our huge brains is discussed in C. Wills, *The Runaway Brain: the evolution of human uniqueness* (New York: Basic Books, 1993).

INDEX